MW00845670

A Viewer's Guide

for

Connections: Technology and Change

by

Kiki Skagen Munshi

with

Jane L. Scheiber

This Viewer's Guide was developed by
Courses by Newspaper, a project of University Extension
University of California, San Diego

Funded by the National Endowment for the Humanities

Boyd & Fraser Publishing Company
San Francisco

Contents

Preface

This Viewer's Guide has been specially prepared to further your enjoyment and understanding of CONNECTIONS: TECHNOLOGY AND CHANGE. America's first national media program to feature both a newspaper and a television series, CONNECTIONS: TECHNOLOGY AND CHANGE was developed by Courses by Newspaper, a project of University Extension, University of California, San Diego, with funding from the National Endowment for the Humanities.

Components of the program include a ten-part television series, "Connections," coproduced by BBC and Time Life Films and featuring writer/narrator James Burke of the BBC; a narrative text, *Connections,* by James Burke, which is the companion volume to the television series; a series of fifteen newspaper articles written for Courses by Newspaper and coordinated by John G. Burke, Professor of History at the University of California, Los Angeles; a Reader/Study Guide, *Technology and Change,* edited by John G. Burke and Marshall C. Eakin; and this Viewer's Guide.

"Connections" was presented for the first time on American television on PBS by WQED/Pittsburgh with a grant from the American Telephone & Telegraph Company and Associated Companies of the Bell System in the fall, 1979.

Designed for both the interested viewer/reader and the student enrolled for credit, this volume relates the video and print components of the multimedia program. Both the television series and the print materials deal with the "connections" inherent in the course of technological development and among the technologies that dominate our world. In addition, the two Burkes were in consultation during the development of the program. Each component, however, presents a different approach to the subject of technology and change. James Burke uses the television component to present a series of detective stories that connect seemingly unrelated events as he traces the evolution of eight modern inventions. He thus presents the historical flow of technological development, and then raises questions about the nature of technology and our dependence upon it from this base. John G. Burke, in contrast, has structured the newspaper series and the book of readings around the questions themselves, focusing on such issues as the effects, preconditions, and sources of technological change.

The aim of this Viewer's Guide is both to highlight important points and themes of the television series and to tie those themes to the different perspectives of the newspaper articles and the articles in the Reader. In addition, this volume provides some interesting background on the filming of the "Connections" programs, which took James Burke and his crew to more than 150 locations in nineteen countries. Also included are questions for discussion and study. Finally, the Time Chart is designed to give the reader a sense of the chronology of the developments discussed in the video and print materials.

Many people share the responsibility for the development of the Viewer's Guide. George Colburn, Project Director of Courses by Newspaper at the University of California, San Diego, has been perhaps the single most instrumental person in bringing the television series and the course by newspaper together. He also shared in the basic conceptualization of the Guide. Elliot Wager of the Courses by Newspaper staff did much of the work on the Time Chart. Both John Burke and James Burke have been kind enough to review the manuscript, and James Burke also contributed the information for the Program Background sections. Thanks are due to all of these, but the greatest debt is owed to Jane Scheiber, whose careful editing and perceptive suggestions have become a substantial part of the final volume. We also wish to acknowledge our gratitude to the National Endowment for the Humanities, which funds Courses by Newspaper, and to its Director of the Office of Special Projects, James Kraft, for his support of this multimedia venture.

Although Courses by Newspaper is a project of the University of California, San Diego, and is supported by the National Endowment for the Humanities, the views expressed in course materials are those of the authors only and do not necessarily reflect those of the funding agency or of the University of California.

Kiki Skagen Munshi

PROGRAM ONE

The Trigger Effect

Program Summary

"Look around you," requests James Burke as he opens the first program in the "Connections" series. Look around—and *see* the technology upon which you are dependent. What does it do to you? What do you do to it? What were its origins? How did it grow? What would happen if it weren't there?

Burke leads us through the light-splashed night of New York as he talks about the technology island we have created on Manhattan—and in other major population centers of the world. The use and understanding of technology do not, he says, go together. The elevator doors close between us and his face, leaving us to watch the gears and pulleys that pull the box up into a skyscraper. Burke thus shows us the reality behind the wood-paneled interior of the movable box. And he uses this example to evoke a fear many of us share about one particular manifestation of technology. What if the elevator stops halfway up and the lights go out?

In November, 1965, Burke tells us, it did stop. New York's vulnerability was graphically illustrated when a rather small piece of equipment did what it was supposed to do. The evening began normally. On the subway, in the airport, in hospitals, on the streets, at the United Nations, and at the headquarters for ConEdison, people followed their routine activities. Leisurely at first, then at an increasing pace, the camera cuts from one scene to another as we approach the point at which Adam Beck II power station failed, draining light and heat from the great city.

The response of those caught in the New York blackout of 1965 was to "make do" with the technology available while they waited for more technology to come to the rescue. But what, Burke asks again, would we do if technology didn't come to the rescue, if it failed altogether?

If it all failed, supplies would run out and we'd have to leave the technological island, the city. Burke strides through a bleak landscape of abandoned cars and deserted freeways, talking about the few alternatives that would then be left for survival. For example, today's farms run on power—gas for tractors, electricity for power saws, drills, sterilizers, light, water, and sewage. The only means for survival would lie in the simple technologies that did not depend on these sources of energy. *If* in a dusty loft we were able to find a plow and harness, *if* we had draft animals, *if* they could be harnessed (and the cows that Burke puts in front of the plow would not normally take kindly to such carrying on), and *if* we had the technical know-how for old-fashioned farming, then and only then might we survive.

The basis of civilization rests on food and the means to produce it. The scene changes to desert and oasis, one of the places where the weather changed about 12,000 years ago. When the grasslands dried into desert, nomad hunters headed for water. In Egypt this was the Nile, and Burke uses the rope and wood of an Egyptian plow to illustrate his point. The plow, according to Burke, was the "first man-made trigger of change," and it allowed humans to predict that there would be food in a particular place at a particular time. The ancient Egyptians could, and had to, plan for the future. They needed to store grain. They needed to cook it. These needs gave rise to the technologies of ovens and potting and baking. Finally, the need to know when the Nile would flood and the need to allocate those flood waters according to a predetermined plan brought about a strong, central governmental system that quickly became an empire.

The same kind of trigger effect—one that changes a nomad society into an empire—is now occurring in the Arab states. The questions it raises about the causes and effects of change are, by and large, the same that can be asked about ancient Egypt and about our own society. We will explore them in this chapter because they form the basic themes of the rest of the series.

Program Background

This program takes us from the top of the World Trade Center in New York to the top of towers in Kuwait as James Burke explains our dependence on modern technology and discusses rapid change in the modern world. Eight hundred thousand people were trapped in the subways in the New York blackout of 1965. The group of them shown in this program have re-enacted their experience, as have the people in the other dramas of that day. These scenes were shot in New York City and Kennedy Airport. The deserted freeway scene was taken on Staten Island—on a section of road not yet open to traffic.

The Egyptian scenes were filmed on location in Egypt, which in many places has changed little since ancient times. The potter makes his living casting pots on that wheel. The step pyramid and tomb are by Cairo, but the great pillared hall is a part of the Palaces of Thebes in Luxor, Upper Egypt. The Arab falconers were, of course, in Kuwait, where the last scenes of the program were also filmed, in a viewing area atop great water towers.

Themes

The many questions James Burke raises during "The Trigger Effect" fall into three general areas. All of them are important to every individual in today's world, whether or not he or she thinks about them.

The complexity and systemic nature of technological society. "Technological society" really includes all societies. Burke illustrated the technological complexity of ancient Egypt and the ways in which people, born over 4,000 years ago, used their tools to build monuments we still marvel at today. Even without those monuments, or the capacity to produce them, ancient societies were still "technological" in a very basic way. They used implements for farming, cooking, and storing food. They made clothes and shoes. They created harnesses for draft animals and constructed buildings in which to live.

As the complexity of technology increases, however, two things happen. First, a greater number of processes come into being for an end product. In ancient Egypt the potter mixed the clay, spun the wheel, threw the clay on it, and formed a pot in which to store food. In contrast, the plastic container in a modern refrigerator is the product of many more—and more intricate—processes that began, in part, with the extraction of oil from the ground.

Second, the user of technology becomes further removed from the processes by which artifacts—the things that are used—are made. Not all people in ancient Egypt could make pots. But most of them could describe how they are made. Very few of us can do the same for plastic containers.

Moreover, none of these technological processes occurs in isolation, for every aspect of technology is an aspect of a whole system. A culture's technology is a structure, composed of many pieces, and all interlock; one part is laid on another part and cemented with a third. An "instance" of technology does not just happen, and this implies something else. If one part of the technology is changed, there is some impact on other parts of that technology. The impact may not always be predictable—all we can definitely say is that, because of the systemic nature of technology, there will be an impact.

The hows and whys of technological change. Burke's basic thesis in "Connections" is that change occurs as a result of many factors, but only under certain conditions. The most important is that a "technological infrastructure" must exist. The Egyptians could not have invented the plow if they had not known how to work with wood or domesticate animals. Second, for the technological change to take hold, be used, and have an effect, there must be a need for it. Pots were not made before there was a surplus of something that people wanted to keep.

Specific reasons that change occurs, however, are more difficult to identify. Burke believes that because of a "trigger effect," change in one area leads to a chain reaction of other changes. What brings the initial change into being? It may be individual ingenuity, an economic need, the result of war, or the outcome of a number of other specific occurrences. We will see several of these chains during the coming programs and, as Burke says, "it will become clear that history is not, as we are so often led to believe, a matter of great men and lonely geniuses pointing the way to the future from their ivory towers. At some point every member of society is involved in the process by which innovation and change comes about. . . ." This brings us to the third theme.

The past, present, and future effects of technology. The opening scenes of the television program demonstrate the tremendous impact of modern technology on the people who live in New York City and on the world they inhabit. Most of them are changed physically by their technological environment. Count the people who wear eyeglasses. Look at their builds. They are tall and well fed. Perhaps some of their muscles are not well developed because they rely on other forms of technology such as the cars, which move in more or less ordered lines.

Those lines, perhaps, are symptomatic of the ways in which technology shapes society. It would be difficult to drive if we could not predict, with reasonable accuracy, the movements of the cars around us. We can, however, in part because we have to make and follow rules about movement that simply wouldn't be necessary in a society where everyone walked. This is a small example, of course. A larger one is implied by the existence of those cars. Someone owns them, someone must keep them running, someone is responsible if one hits and kills another person. The cars exist in more than a technological environment. They must, of necessity, exist in a legal, political, and economic environment as well.

And finally, the cars and all of technology have an impact on the physical environment in which they exist. The island of Manhattan—a great shell of concrete, asphalt, steel and glass dropped over a hump of dirt and rock—may be the ultimate symbol of the way technology reshapes the world we have inherited from Nature.

Readings

Read Chapter One in James Burke's book, *Connections.*

Many of the articles in the Reader, *Technology and Change,* apply to these themes, but they provide different approaches to the questions and different interpretations of events, phenomena, and ideas. James Burke, for instance, used the 1965 blackout in New York to illustrate our dependence on technology. Barry Commoner in "Are We Really in Control?" (pp. 6–9) would, no doubt, agree with the fact of that dependence. But he sees the blackout primarily as illustrating the fact that we are creating technological systems we do not fully understand and cannot fully control. That theme—is there control over technology?—continues to be explored by people with vastly differing ideas: Lewis Mumford, in "The Technique of Total Control" (pp. 10–12), sees technology as a threat to freedom; Jacques Ellul, in "The Technological Order" (pp. 13–19), believes it is no longer possible to take command of technology; Samuel C. Florman, in "In Praise of Technology" (pp. 21–30), takes up and refutes each of these points, arguing that the problem is not technology but our demand for its benefits.

The question of technological determinism—is technology the determining force in human affairs, or are humans themselves responsible for their world—is continued in subsequent articles. Peter F. Drucker takes a middle line in his article "The First Technological Revolution and Its Lessons" (pp. 39–46). He believes that the conditions that technology created determined many social outcomes, but that humans controlled the philosophical and ethical basis of society. Drucker's discussion of irrigation civilizations continues and expands the material from the television program and the book *Connections,* although it differs in a few details. Drucker also discusses some of these questions in the second newspaper article, "Silent Revolutions." Similarly, Jacob Bronowski's selection, "Technology and Culture in Evolution" (pp. 55–59), picks up and expands James Burke's point about the "future" that was created in Egypt with the invention of the plow and the cultivation of crops.

Simon Ramo's selection, "The Systems Approach" (pp. 77–82), uses the complexity and interrelationship of technology and society to illustrate an intellectual tool: analysis and design applied to the whole rather than component parts. With this and with the tools provided by modern technology, he asserts, we have the means to find answers to problems such as those raised by Burke.

John Burke, in "Technology on Trial," the first newspaper article for the course, addresses the themes we identified above, and raises questions in each area. Note the differences between his and James Burke's approach. No fundamental disagreement emerges between this article and the television program, but we could easily imagine the two men debating the importance attached to certain factors. This difference in emphasis (John Burke designed the newspaper course and edited the book of readings) will continue throughout the coming programs, and we will return to it occasionally in the chapters ahead.

Discussion Questions

1. Put in writing your impressions of the way James Burke feels about technology. (A) Put it aside for a few weeks, then read it and see if you've changed your mind; (B) Contrast his ideas to those of either Ellul or Simon.

2. Describe an example in the television program "The Trigger Effect" that you believe illustrates the systems approach to a problem.

3. Do you believe the ancient Egyptians' dependence on technology was qualitatively different from that of the modern New Yorker? If you do, you might try to put these differences in opposing columns, then compare them.

PROGRAM TWO

Death in the Morning

Program Summary

This program is the first of the "detective stories" that will transport us from an event far in the past to an invention that influences the modern world—in this case, the atomic bomb. It opens with Burke on a sailing boat, looking ahead across the Mediterranean. No one, he states, can see into the future. We only guess the effects of our actions.

This particular story begins with a touchstone in an Eastern Mediterranean marketplace some 2500 years ago. The black piece of rock has a polished side that can be marked by gold, and its discovery meant the standardization of coinage with the consequence that trade replaced barter. Trade in turn brought more ships to Mediterranean ports and the dissemination not only of goods, but of knowledge.

The scene changes to the busy markets and flat-topped cement roofs of Alexandria in modern Egypt. Burke, walking among ruins against a white hot sky, talks about the burning of the great library of Alexandria, built in the third and fourth centuries B.C. Not all the books were destroyed, however. He climbs down into underground caverns made for the scrolls and pulls out a chart that was 1400 years ahead of its time: Ptolemy's star tables. It wasn't much use to navigators when it was written because of their sails.

The square sail then in use would sail only with the wind, and if you can go in only one direction, precise navigation is neither practical nor necessary. This type of sail lasted until about the eighth century, when Arab pirates made it smart business to minimize losses by dividing precious cargoes among many small boats. A montage of burning ships, coins and jewels, and attacking men dissolves into an Arab dhow, a small boat with a triangular sail, or lateen. This lateen sail made sailing against the wind possible. Now ships could be sailed at almost any time in almost any direction. Together with the square sail, the lateen was used in sixteenth-century ships like Sir Francis Drake's *Golden Hind*. Standing on the prow, behind the figurehead, Burke tells us how the sails work. Later, in Drake's cabin, he demonstrates the magnetic needle that pointed to the North. Drake's sails and compass enabled him to sail around the world, and he returned with a rich cargo of treasures captured from Spanish ships and settlements in the New World.

The Elizabethans needed accurate navigation for their trade. Ptolemy's star tables were now used by navigators to fix their north-south position, but the companion tool, the compass, was reported to be unreliable. Scenes of the Queen on the Thames, of ships unloaded by torchlight, of gaming and trade lead into Hampton Court, where we find Burke, discoursing on the compass. An Elizabethan doctor, William Gilbert, conceived of the earth as a magnet, and, to support his belief, was able to demonstrate that magnetized metal balls also attracted compass needles. Compass needles, he theorized, must point to the earth's magnetic north pole, not the north star. A by-product of his thesis—that there was a vacuum, an absence of any matter, between planets—led to a series of experiments in Germany that produced another important by-product. In one of the experiments a rotating sulfur ball on a stick glowed in the dark and crackled. It was the accidental discovery of electricity.

According to Burke, a note about the crackle and glow, buried in a treatise on another matter, could have led down a number of different paths. The route he chooses takes us through an exploration of lightning and weather, by means of a number of vehicles. Burke sails above green fields in a modern balloon, similar to those used in the nineteenth century for weather observations. Then we see a party of Victorian enthusiasts travel by foot and horseback to the top of Ben Nevis, a mountain in Scotland, to inaugurate a weather station that was ultimately responsible for the creation of a cloud chamber. The end of this journey of inquiry is the cockpit of a modern vehicle—a B-29 bomber from World War II. These planes were safer, and we are safer as jet passengers today, because of one result of the journey we have taken with Burke—radar. But all of us are in grave danger because of another result. The camera moves away from the plane to reveal the name: The Enola Gay. The Enola Gay dropped the atom bomb on Hiroshima one death-filled morning in 1945.

Program Background

This program shows Burke on land, sea, and in the air as he recreates events over a 2500-year period. The ship scenes for "Death in the Morning" use a variety of sites and techniques. The opening shots show Burke in Alexandria harbor in Egypt. The pirate attack sequence, however, is a set of visual effects, created in the BBC studios. Exeter Maritime Museum in England is the home of the two-masted Arab dhow that Burke demonstrates, and the *Golden Hind* is located in Brixham, also in England.

The demonstration of Gilbert's hypothetical "vacuum" was recreated in Regensburg, Germany, during the annual Bierfest, with period costumes and the same results Gilbert supposedly obtained. Period costumes also highlight the recreation of the arrival of Elizabeth I and her entourage at Hampton Court on the Thames River, outside London. The recreation of the opening of Ben Nevis Observatory is, again, on location in Scotland, while the closing sequence of the film was shot in Harlingen, Texas, home of the only B-29 still in flying condition.

Themes

One of the most elusive relationships in technology, yet one of the most powerful, is the one between technology and knowledge. Knowledge about technology and the ability to use it is power and, in many circumstances, the means to wealth. Commonly accepted "knowledge" based on technology can change social institutions and the further growth of technology. Knowledge and technology act together, each one spurring individuals to discover more about both and each shaping the way technology grows. Again and again in these programs we shall see accumulated bits and pieces of knowledge come together to produce a technological leap that sets in motion a complex train of further changes.

Common "knowledge" and the growth of technology. Knowledge about the touchstone provided an impetus to its use. And the "knowledge"—of the purity of gold—that the touchstone made possible increased trade and led to changes in the design of ships.

Barter is a particularly awkward way to transact business. Imagine each owner of a stall in the Alexandria market having to exchange tomatoes or pots directly for everything his or her family needed!

The touchstone made possible a common medium of exchange. Many items—for example, shells—were used for money before it appeared. But if various pieces of "money" differ from one another and traders argue about the value of their money as well as the value of their goods, the business of business is uncertain and slow. Gold bullion probably existed before the touchstone came into use, but its metal was often adulterated. The touchstone was a means of standardization that was commonly accepted. It therefore provided the impetus to trade in goods and knowledge that Burke uses on the next step in his journey.

Knowledge and power. As the technology of trade and commerce grew through the centuries, so did the *value* of knowledge concerning it. Those who possessed charts or knew how to use the sky tables possessed rare and precious information. They possessed a "good" that was as valuable or more valuable than the objects they carried. They could sell their knowledge or use it to other ends because knowledge, coupled with the means to use it, is power. But power, of course, leads to more than the simple accumulation of wealth. It means a special place in the social hierarchy and, ultimately, the ability to rule. It also means the ability to destroy. And both power and destruction change the one who acts as well as the one who is acted upon. The impact of the name, the Enola Gay, is not simply a result of the plane's role in dropping the first atomic bomb. The bomber is a symbol of the great questions about the uses of technology, knowledge, and their powerful effects on all of us.

The integration of knowledge and technology. The desire for knowledge and the need for knowledge in order to use technology are illustrated at several points in this program: the use of the compass and the search for explanations of its inaccuracies; the trips in balloons and up Ben Nevis to discover more about the weather; and Charles Wilson's attempt to explain the "glory" he saw in the Scottish sky.

These examples, however, also illustrate a factor in the Western (or European) search for knowledge that has been of particular importance in shaping our technology. This is a desire for control over our environment or (if we can't control it) the ability to predict accurately what it will do. Reliability, predictability, control—all form the underpinnings of Western technology.

But knowledge and technology do not always accompany one another. Ptolemy's star tables were not coupled with maritime technology until centuries after he compiled them; the compass was used before the principles on which it worked were fully understood. In this and in other programs, though, we shall see that, in the majority of cases, when a technological process was used there was almost always someone who wanted to find out why it worked as it did. The result was such an intermingling of science and technology that it is difficult for us, today, to consider one without the other.

Technology and the economy. Another theme in "Death in the Morning" is the double relationship that exists between technology and economic gain. The touchstone, the lateen sail, the compass, and accurate weather prediction increased the ability to conduct trade and commerce—and thus wealth—but each was developed or "discovered" in order to increase potential profits. The interaction between technology and the profit motive will appear, in one form or another, in most programs in this series and will be discussed at some length in connection with Program Five.

Readings

Read Chapter Two in James Burke's book, *Connections.*

One of the most powerful tools (and the product of very advanced technology) to assess information and enhance knowledge is the computer. Its possible effects on our society can be seen in several different ways. To Herbert Simon, author of the Reader selection "What Computers Mean for Man and Society" (pp. 68–76), computers make information more accessible and more easily used for the benefit of the general public. But the most important question he poses goes beyond the "mechanics" of computer use. It concerns our perception of ourselves and our place in the universe. Imagine living in the early sixteenth century and being told that the earth, which you know is flat because you have traveled on it, is really a sphere. The computer may have a similar effect on all of us before long as it is increasingly able to duplicate functions of the human brain.

Lewis Mumford, on the other hand, believes that the avenues opened by computers lead towards control by the few. In "The All-Seeing Eye" (pp. 65–67), he argues that the knowledge available through technology will, by means of technology, be used to shape not only the environment but most of us living in it. To both Mumford and Simon, knowledge—whether in the form of the navigational charts we saw in the program or of computer print-outs—is power. The key variable is in the way it will be used and how it will affect us.

Marshall McLuhan also sees "another way of thinking" in modern technology, but adds a further dimension to the question of control. In his article on "Automation" (pp. 110–113), he contends that our growing ability to regulate work-processes through technological means will increase the distance between modern men and women and the brutal, tedious lives of their ancestors. The question of the "good life" and technology's impact on it will be presented by James Burke in a later television program; McLuhan's article is a prelude to that as well as an illustration of the whys and wherefores we find in our search for knowledge and control.

The systemic nature of technology will also continue to be illustrated throughout this course as we meet ideas, people, and innovations over and over again in different contexts. Derek de Solla Price, in the third newspaper article, "How Terribly Technical!," identifies developments in what he terms "an aggravation of the elitism of knowledge skills," which we shall encounter in subsequent programs. One is the Gutenberg Revolution, caused by the invention of movable type (Program Four), and the second is the Scientific Revolution (parts of which will appear in most programs). The increasing specialization of technology identified by Price is the reason that Sanford Lakoff, a political scientist, believes the role of government in the advancement of technology is crucial. In the selection "Knowledge, Power, and Social Purpose" (pp. 68–71), Lakoff argues that we also need an increased understanding of the relationship of government and the growth of scientific knowledge in order to prevent a loss of control over how knowledge is used and an abdication of the role we, the people, play in shaping our lives.

Discussion Questions

1. What was the importance of the touchstone?

2. Simon states that "energy and information are two basic currencies of organic and social systems. A new technology that alters the terms on which one or the other of these is available to a system can work on it the most profound changes." Find an illustration of this statement in the television program "Death in the Morning," and explain your reasons for choosing it.

3. The reason the horses were unable to pull apart the two hemispheres in Guericke's experiment is based on a simple physical principle. Explain it or find an explanation of it. What was its significance in the program?

4. Find instances of the need to regulate or predict environmental events in both the television programs and the readings to date.

5. What are your own feelings about computers? In what ways do you relate to them in your everyday life? Start with the pocket calculator.

PROGRAM THREE

Distant Voices

"Distant Voices" opens with Burke carrying an innocuous suitcase that is made sinister by its contents—a nuclear bomb. Its existence, Burke tells us, is as unbalancing to the present military situation as another invention, the stirrup, was in its day. The camera pulls back to show Burke in a green field, an abbey in the background. The field is Hastings, on the Southeastern Coast of England, where William of Normandy (later known as the Conqueror) defeated Harold, Saxon King of England, in 1066.

In the famous Bayeux tapestry, celebrating the Battle of Hastings, we see the shield and lance whose use depended on the stirrup. It was this device, according to Burke, that gave the Normans the advantage; the result was a change in the rulers of England. A slow-motion sequence illustrates this means of fighting and its development after Hastings. In a medieval tournament, amidst jousting, eating, laughing and looking, the winner took all, including the opponent's horse. This was, at least until knighthood was made hereditary, a way to social respectability. But despite the circus-like atmosphere, the local tournament was serious business, and the knight was a powerful and frightening war machine.

The knight's supremacy ended in France in 1415. On camera, Burke leans over the black entombed figure of the English King Henry V as he describes the bloody encounter with the French in that year at the Battle of Agincourt. On this sea of mud Henry's Welsh archers loosed their arrows from a powerful new weapon, the longbow, to bring down horses and men in a frenzy of slaughter. The simple longbow conquered the knight and his way of fighting after some 300 years of power.

Three other inventions, developed before Agincourt, were in the long run to bring about the demise of the longbow. Hand-painted illuminations of plowing, sowing, and threshing introduce this segment of the program. The first invention was the moldboard plow. Maneuvering a medieval wood plow with its vertical knife, Burke plows up the field of Agincourt. The next two inventions were the horse collar, so the animal could pull the plow without choking, and the horseshoe. A blacksmith taps a shoe on a horse's hoof as Burke explains the role horseshoes played in transporting the new surplus of food that resulted from better plows and crop rotation. People now had money to spend on weekly fairs and markets with their baths, amusements, goods, medical care, and other inventions—and they didn't practice the longbow. The King tried to ban these leisure activities, but a new way of killing people appeared in the nick of time.

Burke continues his story from a pig sty. The pigs' manure and urine formed a basic ingredient for gunpowder. The picture cuts to China where gunpowder was used for fireworks in religious rituals, illustrated by a swirling dragon and praying girls. Burke walks into the temple to talk of the reasons the Chinese invented so much yet used so little for change. This sprang, he asserts, from two sources: from the Chinese view of life; and from the fixed society and bureaucracy that were a result of the need to administer the ancient, complex, irrigation system.

It took a somewhat more open system, such as that which existed in the West, responsive to the profit motive, to do things like put gunpowder in a bell and make a cannon. As explosions take place, we move from the bells and a bell-foundry to a small medieval town in Italy where the first cannons were used, then to a town in the forests and mountains of Czechoslovakia where silver was mined to finance them.

The problem of raising water out of mines such as those brings Burke to the next invention. Puzzled as to why water couldn't be lifted more than 32 feet, someone wrote to the eminent scientist Galileo, who wrote to his assistant Torricelli, who wrote to his colleague Ricci, who sent the letter to France. Burke, in a modern yellow French postal van, explains the importance of communications in putting pieces together and getting them to the inventor and mathematician, Blaise Pascal. Pascal used mercury to make a small model of the water pump. Then—the picture tilts and the van crawls up the screen—his brother-in-law went up a mountain to test atmospheric pressure at different altitudes. Burke repeats the test, illustrating the creation of the barometer.

There are many ways to go from this point. Burke decides to follow communication. He leads us to a Frenchman carrying a barometer affected by electricity, through a series of interesting (often silly) experiments with electricity, to a succession of other important developments, including the demonstration of the link between electricity and magnetism. These came together in the late nineteenth century. Electricity created a magnetic field that could be used to transmit vibrations from sound along a wire. The result was the telephone, "invented" (because of a head start to the patent office) by Alexander Graham Bell. Today, we have complex telecommunication systems such as the one Burke shows us in Puerto Rico that enable us to communicate with the rest of the universe.

Program Background

The land Burke plows as he talks about agricultural developments is the battlefield of Agincourt. The plow was specially reconstructed for this sequence, but other artifacts in the program are originals. The sword Burke holds in Westminster Abbey, for example, was Henry V's, and the embroidered pictures depicting stirrups and lances are from the Bayeux Tapestry in northwestern France. The tournament and market scenes were both recreated for the series, the first at Ludlow Castle in England (where Mark Wing-Davey delivers the Crispin's Day speech from Shakespeare's *Henry V*), and the latter at Grandson, Switzerland. Other locations for this program include Taiwan; Brittany, France; Cividale del Friuli, Italy; Jachymov, Czechoslovakia; and Puy-de-Dôme, France. The final sequence was filmed on location in Arecibo, Puerto Rico.

Themes

Culture, society, and technology. A major theme throughout "Distant Voices" is the interaction between culture and society, on the one hand, and technology on the other. Some portions of the program imply that the development of technology depends upon the culture and society in which it occurs. Others point out technological changes that determined the shape of society. Perhaps the only possible conclusion is that the two interact, and that the question of which is the dominant factor depends on many other things.

Much of the program shows the impact of technology on society. The use of the stirrup and lance and the rise of the knight with his expensive horse and equipment put war into the hands of an elite class. The distance between a man on a horse and one on foot, after all, is more than the few feet of height given the rider. It is a distance created by the territory the horse can cover, by the cost of the animal, and by the psychological advantage given a rider in a world where horses are the fastest and most powerful form of transportation. All of these things helped restructure society and the culture that accompanied it in the twelfth and thirteenth centuries.

About the same time, other far-reaching changes had been set in motion by the invention of the moldboard plow, the horse collar, the horseshoe, and the three-field system of crop rotation. It took longer for their effects to become cumulative, but the changes they wrought were longer lived than the knight and his world—the creation of an agricultural surplus, the growth of markets, a rise in the standard of living of many people, and a consequent growth in commerce and communications.

The importance of the impact of culture and society on technology is most obvious in the Chinese sequence. Burke uses the invention of gunpowder as a springboard to ask why the Chinese invented so much. The Taoists, he answers, were interested in discovering order in the universe and, therefore, spent time investigating it. Why then, he continues, didn't the Chinese put their inventions to practical use? Burke offers two reasons: One is that the Chinese were prone to regard the universe as fixed rather than changeable. The environment was a given, to be lived in, rather than a tool to be altered and used. The second reason sprang from the great irrigation systems of the Chinese river plains. These required an administrative bureaucracy that dammed the flow of social change.

Another set of reasons appears in later programs. They have to do with attitudes toward certain kinds of work.

In Chinese society, as in most European societies, hard manual labor was reserved as much as possible for the lower classes. The Chinese, however, were more rigid than the Europeans in maintaining class distinctions. The result was that the people who thought about things and the people who used them physically were different. The impetus to "make things a little easier" and to "make things a little more efficient" never took hold in the same way it did in the West.

The cumulative result was a society that pursued many areas of inquiry but applied few of them to practical matters. China was not unique. Many other Asian societies were similar, and we will see in "The Long Chain" that some of the same arguments can be used to explain the difference between American or British and German inventiveness in the nineteenth century.

War and technology. In the West, war was a tremendous impetus to the development of technology and changes in the social structure. We see several examples in "Distant Voices": the use of gunpowder in cannon; the formation of a new social class through the use of the lance and horse. The reciprocal influence, of society on war, is demonstrated by the longbow's demise. These are themes that will recur again, most specifically in Programs Eight and Nine, and that will be discussed in Chapter Eight of *Connections.*

Readings

Chapter Three in James Burke's book *Connections* corresponds to this program.

Four articles in the book of readings, *Technology and Change,* deal directly with the interaction of culture and technology. Robert Heilbroner, in "Do

Machines Make History?" (pp. 145–153), and William F. Ogburn, in "Technology as Environment" (pp. 154–160), both believe that, with certain qualifications, technology does shape society. George H. Daniels, on the other hand, argues in "Technological Change and Social Change" (pp. 161–167) that social needs shape technological innovations. In the next article (pp. 168–174), John A. Hostetler describes the interaction of technology and society in a particular instance, the Amish, whose values kept them from adopting modern technology.

The selection by Clarence Glacken, "Nature and Culture in Western Thought" (pp. 121–125), presents the European image of man's place in the universe. It is a central and an instrumental spot. One aspect of this is that the environment shapes culture but humans, in turn, shape the environment to their ends. More central to our purposes, perhaps, is that the importance of man in relation to other beings and objects is greater in Western thought than in many other cultures. Man is not a part of an intricate whole, occupying a space about equal to his size. He is of more consequence than anything else. The reason for being revolves around him and he, therefore, is empowered to change the rest of the world to suit his ends.

This image of man is reduced by Jacques Ellul, whose view in "The Technological Order" (pp. 13–20) is that society is irrevocably shaped by technology. It is hard to imagine Ellul saying there was any time when technology had *no* effect on society, but in today's world, technology has taken control of us. Nathan Rosenberg, on the other hand, sees technology as heavily influenced by both society and resources. In "Technology and Resource Endowment" (pp. 126–129), he states that "... resources establish the particular framework of problems, of constraints, and opportunities, to which technological change is the occasional human response.... Technological change... does not *occur* in the abstract but rather in very specific historical contexts. It occurs, that is, as a successful solution to a particular problem thrown up in a particular resource context."

You might also want to read Lynn White, jr.'s article, "The Act of Invention" (pp. 382–383). He presents an account of the stirrup and lance that varies slightly from Burke's.

Two of the newspaper articles for the course relate closely to this program. The second article, "Silent Revolutions" by Peter F. Drucker, describes the influence of technology first on the role of women, then on the ancient Egyptian society we saw in the first television program. Drucker argues that technology and society change together, and the ways in which they change must be compatible or change will not occur. Like James Burke, he uses China as an example of an alternate mode of approaching the environment and change.

Somewhat akin to Drucker, Edwin T. Layton, Jr., argues in the sixth newspaper article, "The Influence of Societal Values," that "there is no inevitable cause-and-effect relationship between technological and social change."

Discussion Questions

1. Do Burke's views in the program "Distant Voices" correspond more closely with those of Heilbroner or Daniels? Explain your response.

2. Glacken's Reader selection presents three ideas that, he believes, have influenced Western thought and culture from ancient times to the present. Do you agree with him? How do they fit with Burke's—or another source's—view of Chinese culture and civilization?

3. A television program often imposes on events a linear sequence—first, second, third, fourth—that doesn't, in fact, exist. Construct a simple timeline for this program, based on your recollections; then compare it with the Time Chart on pp. 61–64 of this Viewer's Guide.

PROGRAM FOUR

Faith in Numbers

Program Summary

Kneeling in weeds against a stone wall, Burke explains how satellites and computers can pinpoint a location anyplace on earth. He checks the longitude and latitude numbers he is given by the computer on a map and stands up. He is at the site identified through the satellite—the Barbegal Aqueduct in France.

Our world is held together by the kinds of organization and communication demonstrated by the satellite and computer. "But what will that organizational network do to us next?" Burke asks. Perhaps, he says, we can learn from the Roman Empire because a vast, highly organized network of communications held a world together before, and then collapsed.

In the Middle Ages, the one organization that functioned internationally was the Church. Through its communication system the knowledge lost as a result of the fall of Rome gradually spread again. Part of this knowledge was an understanding of how water mills and their gearing systems worked. Burke shows us a wooden model of a wheel which powers, in turn, a grinding mechanism, a trip hammer for a suction pump, a bellows, and a crank for a sawmill. These systems and water power led to a "Medieval Industrial Revolution"—a spurt of productive activity occurring between the tenth and fourteenth centuries.

This revolution was led by monks. The Cistercians with their white robes and black hoods glorified work. They chopped wood, hoed, made wine, herded sheep, and told one another across Europe how they did it. Sheep-rearing may have been their biggest success, and the wool they produced could be made into cloth for all Europe. But they needed a more efficient loom and a spinning wheel to provide it with yarn. Both came from China in the fourteenth century.

Banners and the stone buildings of Bruges provide a backdrop for a procession of medieval guilds. The city was made rich by cloth. The whole world traded its goods for the woolens of Flanders.

Much of the trading was done at the Champagne fairs in the South of France. The fairs were supported by governments, by the merchants and, in an important development, by the new investment contracts that allowed stay-at-homes to participate in the benefits of trade. Commerce flourished.

But the fairs and their trade were destroyed in the mid-fourteenth century by an invisible enemy. Scenes of the busy medieval market give way to a deserted lane, rain, and dead rats floating in a gutter. A solitary man with a horse cart paints crosses on doors to mark the Black Death. Half the population of Europe fell victim to the bubonic plague.

Amidst all the grimness, however, was one brighter note: when the Black Death receded a few years later, most of the property was left to fewer people. Their race to grab the best of life meant a demand for new luxuries. One of these was linen bedding and linen underwear, used for the first time by common folk.

Linen eventually wore out, and the vast quantity of used linen started a new industry—papermaking. Now there was an abundance of paper but, because of the Black Death, a shortage of scribes. The imbalance was set right by the printing press, which used movable type made in a mold by Johann Gutenberg. Burke demonstrates how molten lead is carefully poured into molds, sets, and is removed. The German "A" he lifts out symbolizes the beginning of one of the most profound revolutions in human history.

The knowledge made available through printing spread with the help of a Venetian, Aldus Manutius. This printer made the world's first pocket editions and brought ancient knowledge to the Renaissance. The Greek and Roman knowledge showed up in many ways, including fountains and automated toys. But automatic mechanical devices were also put to more practical use, such as looms for textiles with new Chinese patterns. Boys pulling strings on the looms were replaced by punched cards that set off the right combinations of threads at the right times. It was a principle that was originally adapted from the cam we saw at the beginning of the program.

The same principle was applied in other areas, such as the riveting machines used in shipbuilding. These ships transported packed masses of nineteenth-century immigrants to the United States. Burke paces the empty corridors and walkways of Ellis Island in New York harbor and talks about the problems such numbers posed for the U.S. census. How do you count them?

By 1890, the question was answered—by punching information on cards adapted from the ones used in looms and shipbuilding. That information was then recorded and sorted mechanically. Do the cards look familiar? They should. Similar cards, working on the same principle, are now used on the invention that runs the modern world—the computer.

Program Background

"Faith in Numbers" was filmed on location throughout Europe. The succession of water wheels came from England, Wales, France, and Switzerland. The Champagne Fair was recreated in Troyes, France, and the old town there was also the scene of the Black Death sequence. The Cistercian Abbey scenes were actually from two monasteries, the Casamari Abbey in Italy and Clos Vougeot Chateau in France. Although the procession of medieval guilds is held annually in Bruges, Belgium, the demonstration of the horizontal loom and the spinning wheel was filmed in St. Luc, Switzerland.

The playful uses of mechanical power—the tortoise and the water-fountained dining room—are from the Palazzo Doria in Rome and the Hellbrun Castle in Salzburg. The Jacquard looms are actually located in a factory in Lyon, France. Another working "antique" that Burke demonstrates is the original Hollerith tabulator used on Ellis Island that is now in the IBM archives in a New York City suburb.

Themes

Technology and work. More than anything else, perhaps, technology is related to work. This fact is woven through the story of Program Four, "Faith in Numbers," and will be taken up again in Program Five.

The theme of work begins in the fields and refectories of the Cistercian monks. Work with the hands was a part of their lives, in tandem with the study and contemplation more usually associated with monasteries. Through this combination, the monks developed new ways of farming and production. And through the network of communication formed by the Church, they made their discoveries known. St. Benedict's rule—"to work is to pray"—destroyed the traditional division between labor and thought and in the process gave to work a nobility and meaning that has remained with us to this day.

But the work was not valued simply in itself. It was also valued for its products—the cloth that people wore, the wine they drank, the books from which they read. The two things together, a belief in the dignity and inspiration of work and a desire on the part of most people for the products of that labor, led to the cumulative growth in the interaction of work and technology that Burke calls the Medieval Industrial Revolution.

The history of work and technology has been, by and large, a move away from the reliance on unaided human strength as a source of power. Sometimes, as with the early Cistercians and the society of their time, the move was largely to an alternate source of power. Far better to build a mill powered by water than to pound cloth and grind grain by hand. But it is also possible to increase the benefit derived from a given source of power without having to increase the power itself. If, for instance, power comes from human action, you can either increase that person's skill and, therefore, his efficiency, or build a machine that augments the force exerted. The foot treadle loom and spinning wheel shown in the program are such machines. They took no more power than the old looms and spindles, but they translated it differently. The result was a lot more cloth and an easier time making it—*if* you were a skilled operator.

In other cases technology does not simply make work more efficient or easier. It changes it completely. The transition from scribe to printing press, as demonstrated in the program, is an excellent example. The labor of the men who set type and operated the new printing press was different in kind and quality from the labor of a scribe. A scribe might be hesitant to accept a press operator's job, yet delighted to be able to buy a cheap, printed book—illustrating the complexity of reactions this kind of change can cause.

Another kind of change in work caused by technology can be unemployment. The Jacquard type of loom wasn't used until years after its discovery for many of the same reasons automation is resisted today by workers who are afraid of losing their jobs.

This program presents several of the factors that bring about developments in labor-saving technology. The influence wrought by the combination of labor and thought of the Cistercians was powerful. A demand for more and better cloth brought successive developments in the loom—but so did the introduction of technology from outside, whether "outside" was China or an organ loft. An abundance of one commodity—such as paper—and a shortage in a complementary commodity—such as scribes—resulted once again in technological change. And, finally, the need for a new kind of job to be performed brings changes. When cards were made to count the growing millions of a young country, they set in motion a train of further events that continues today.

Readings

James Burke provides detailed illustrations of mills, looms, and the printing press in Chapter Four of the *Connections* book that may clarify some of the material in the television program.

The four selections on pages 92–113 of the Courses by Newspaper Reader are particularly applicable to this program. Lynn White, jr., discusses the Benedictines and Cistercians in his essay, "Dynamo and Virgin Reconsidered." He sees much of the progress during the Middle Ages as a result of their influence and of the fact that labor was now being performed by free men rather than slaves. White also talks about the development of the water wheel, adding some information to the facts provided by James Burke.

The next three articles in the Reader deal with the logical outcomes of applying technology to work. What *does* happen? In "Scientific Management and the Assembly Line," Robert H. Guest describes what occurred when Frederick Taylor began to look at people the way people had already looked at machines—a concept that will be further illustrated in Program Five. Guest believes there is merit in Taylor's approach, although it should not be carried to extremes. The next selection, "Worker Alienation," presents some workers' attitudes towards jobs that, because of technology, divorce them from pride in their work. The two other selections on this subject in the Reader illustrate varying perceptions of the impact of technology on work. Siegfried Giedion describes the very real effects of the small electric motor in lightening the physical load of housewives in "Engineering the Household" (pp. 273–275). Ruth Schwartz Cowan, in "The 'Industrial Revolution' in the Home" (pp. 276–282), does not dispute this change, but she believes that the middle-class housewife actually spends more hours completing her "duties" now than she did before these "labor-saving" devices appeared. Cowan argues that more duties were added as the housewife's initial duties became lighter.

The growth of the modern factory system and the effects on workers are described in the fourth newspaper article, "Occupational Destinies" by Joseph Gies. Keep this account in mind as you view the next "Connections" program.

Discussion Questions

1. Identify instances of labor-saving technology in Program Four and discuss them in terms of the articles on worker alienation and automation in the Reader.

2. Lynn White's article presents the hypothesis that the Middle Ages looked to both the Virgin and the Dynamo for their inspiration. Does the treatment of this period in "Faith in Numbers" support his thesis?

3. What was the importance of the cam? How did it operate and what examples of its use are presented throughout the television programs?

PROGRAM FIVE

The Wheel of Fortune

Program Summary

A computer card ended the last program and begins this one. James Burke draws it back from the camera to state, "Computers contain the future within them." The title of the program gives way to a gracefully rotating piece of machinery, moving in time to a Viennese waltz.

The computer Burke shows us runs the planetarium in Rochester, N.Y., which can reproduce the heavens from the beginning of time to the furthest reaches of our imaginations. It is the culmination of a process begun over 3,000 years ago as priest-astronomers watched the moon to tell farmers when to plant their crops.

By approximately 500 B.C. star-gazers had identified the twelve constellations that turn on the screen as Burke names the houses of the zodiac. By the second century A.D. astronomers were able to produce material like Ptolemy's star tables, to which Burke introduced us in Program Two. This time we find them being used in the Arab world.

The Arab universe was composed of seven crystal spheres, and the Arabs learned to predict where, in those spheres, stars and planets would be at any given time. The astrolabe, which they invented to do it, was vital because it told them the direction of Mecca, and what time it was so they would know when to pray.

But such predictions can also be applied to astrology and, through astrology, to medicine. When the Caliph of Baghdad was sick in the eighth century A.D., the monks of Jundi Shahpur, where Ptolemy's tables were preserved, came to cure him. The monks brought a treasure of Greek knowledge and the Arabs took it with them as Islam surged across Africa and Asia into Europe.

Medicine as well as astronomy was popular with the Europeans, and by the thirteenth century both led to scientific investigation. Illuminated manuscripts on the screen dissolve into a tribunal of monks listening to a Latin dialogue on faith and reason. The response of the church to scientific inquiry was the Inquisition, although it found some of the new knowledge useful. Christians, like Muslims, needed to know the correct time to pray. They had sun dials and candles, but sun dials didn't work in cloudy weather and candles were expensive. Even water clocks froze in winter. Not until a twelfth-century invention that got rid of water and used weights was the problem solved. The teeth of gears click past a metal block as Burke describes the verge and foliot system that told people when to pray and when to work.

Other developments soon followed. Springs and their regulation by the fusee were the beginnings of the small watch made in the metal-working town of Nuremberg, Germany, but clocks still were not accurate enough for astronomers. Galileo, who discovered Jupiter's moons with the telescope, was the source of the answer to the astronomers' problem when he wrote about the pendulum. Pendulums regulated clocks more accurately than anything had before for everyone but sailors. Can you imagine a pendulum on a tossing ship?

Navigating east and west across the globe requires measurement of both the heavens and the time. We see the globe spin and dissolve into a planetarium, then the camera descends to show Burke, seated in a glass-making furnace in Sheffield. The glass furnace provided a means to heat steel to the melting point, and in eighteenth-century England this inspired another step in the measurement of time and distance. A good sextant needs accurate scales, and the fine steel from the new refineries was used to make a screw that regulated measurement and marking. But more than a sextant came from this process. The first machine tool had been made. Shots of ships and pulleys, then of a dusty factory illustrate Burke's explanation that machines could now do some of the operations that people used to do. Organized in a slightly different manner, a combination of these operations gave birth to the assembly line.

In the nineteenth century, the Americans took this idea and added a French one—interchangeable parts—to it. The rifle Burke holds is a product and example of both processes—the assembly line used machines that did skilled work rather than people; it produced identical, interchangeable parts.

The result? American industry boomed. Everything seemed perfect except people, and two psychologists set themselves to systematize them. The time and motion studies done by the Gilbreths as shown in an old film completed the development of the modern production line.

But what happened to individuality, Burke queries, as the program ends.

Program Background

In what may be a "first" for television, the water clock in "The Wheel of Fortune" was specially reconstructed from the directions on a fragment of manuscript discovered in the Pyrenees. Also specially reconstructed for this program is the mechanical model of the universe with which Burke explains the Arabs' use of astronomy.

Other sequences in the program are filmed on location or use original artifacts. For example, the debate about faith and knowledge was filmed in the Casamari Abbey in Italy, although the words of the debate were recreated for this program. The clock sequences are from the Strasbourg Cathedral in France. The glassmaking cone, the Catcliffe Cone, is in Sheffield, England, and Jesse Ramsden's "dividing engine" is in the Science Museum in London. The blocks and pulleys used to demonstrate the need for mass production are on the H.M.S. *Victory,* anchored in Portsmouth, England, which is also the location of the block mills that house the original machinery used in Maudslay and Brunel's production line. The film of the Gilbreths' experiment is from the archives of the Gilbreth Estate, while the modern production line in the final sequence is located in Richmond, Virginia.

Themes

Of money and time. For more than 100 years now, the theory of economic determinism has been popular. That is, along with Marx, many of us, much of the time, have believed that our need for money and the things we do to get it have governed the way we live and the course of history. Program Five, "The Wheel of Fortune," is a good base from which to explore that view as it is related to technology.

Which comes first, the development of a technology or the need to use it? Burke might answer that technology does develop, independent of a need to use it, but that it *won't* be used widely until the need appears.

Why? Well, why should I spend time making a telescope if I, personally, have no use for it, or if no one else will pay me for it? Why should I invent a better way to make steel if there were no need, or no market for it among those who make springs for clocks? Man and the products of his imagination, it could be argued, are shaped by a profit motive. Look at the assembly line. The goal was purely economic: to increase the ease and efficiency of production and to produce goods in large quantities.

On the other hand, although a need may exist, the combination of factors that provide its resolution may have nothing to do with economics. And the profit motive, no matter how strong, may not be enough by itself to bring about innovation. In the same example, Burke states that Huntsman's improvements in steel-making were the result of a number of things, the most important being the place he happened to live. Because Huntsman was able to observe the use of clay to line the Sheffield furnaces, he was able to adapt that principle to melting steel.

There is probably no final answer about the exact role of the economic impetus to technological development; it may be that judgments have to be made on a case-by-case basis. In any event, those judgments must be placed in a context with other factors, including the reciprocal influence of technology on economic development. "The Wheel of Fortune" raises the issue of the impact of assembly-line production, and we have seen other examples, most notably in Program Two. And both these questions—"Is technology governed by economics?" and "What is the role of technological change in economic growth?"—are something to consider in coming programs.

Another conundrum is time. The heavens govern time and time governs us. Or does it?

The use of an observatory and the Muslim firmament as an introduction to the invention of clocks was no accident. As Burke points out, the initial observations of the sky were by priest-astronomers who needed to tell farmers when to plant. The rhythm of movement in that sky gives shape to the seasons of our lives.

According to Burke, religious needs initially made the definition of time periods important. The Arabs perfected the astrolabe to identify the direction of Mecca, the Muslims' holy city, and the hours at which to pray, which are set in reference to sunrise and sunset rather than in measured divisions of the sum of day and night. The astrolabe, which told the date and the angle between the horizon and a celestial body, was uniquely suited to this need.

The concept of dividing the day into equal periods was known to the Muslims, of course, but it was the Christians who regularized those periods and made them the basis of activity. This occurred over a number of centuries and, as it was done, it produced many results.

One result was the application of time to daily life. The ability to measure time resulted in the measurement of more things than, perhaps, had been anticipated. These included, most particularly, work. As the change from measuring labor in terms of tasks completed to measuring it in terms of hours spent spread, so did a transition from subjective time (how long you think it takes to do something) to objective time (what the clock says).

The culmination of one use of time is illustrated in the work of the Gilbreths. In their system, time and motion studies change more than the character of work. They change our perception of the nature of the individual doing it as well.

Readings

In Chapter Five of his book *Connections,* James Burke explains some of the more technical material on the television program. Note the different examples the text uses in its description of the development of the modern factory in the United States.

The effect of economic factors on the Industrial Revolution—that technologically rich period of time immediately preceding and following the turn of the nineteenth century when Britain became an industrial power—is discussed in the Reader, *Technology and Change,* by T. S. Ashton. In his selection, "The Industrial Revolution" (pp. 47–54), Ashton describes the ways in which social, legal, and demographic changes fostered investment and economic expansion. This expansion, in turn, provided a context for much of the technological development we see in the "Connections" programs.

Not all factors, of course, are measurable. Some are attitudinal and philosophical. Hugo A. Meier, in "Technology and Democracy, 1800–1860" (pp. 211–213), discusses some of the attitudes Americans held toward technology and growth during the years of great expansion in the early nineteenth century.

The need for economic investment to foster science and the contribution of science and technology, in turn, to economic development are addressed by S. Husain Zaheer in "India's Need for Advanced Science and Technology" (pp. 193–197). Jacob Schmookler, in a different but not incompatible approach, argues that "the incentive to make an invention, like the incentive to produce any other good, is affected by the excess of expected returns over expected costs." His article is titled "Economic Sources of Inventive Activity" (pp. 396–399).

Two newspaper articles are particularly relevant to this program. Gies's "Occupational Destinies," the fourth in the series, applies, as we noted in the last chapter, to this program as well as to Program Four. The eighth newspaper article, "Incentives for Innovation: Technology and the Economy" by Nathan Rosenberg, presents the relationship of technological innovation and capitalist institutions. In it Rosenberg also raises the concept of social costs to a society that do not enter the calculations of an individual company or firm.

Clocks, which occupy a major portion of "The Wheel of Fortune," are also the subject of the Reader selection by Carlo M. Cipolla, "Clocks and Culture" (pp. 60–64). In this excerpt, Cipolla gives more attention to the social context of clocks and time than Burke and less to the technology. His article will be useful in clarifying your sense of the time period Burke has covered.

Discussion Questions

1. *a.* Imagine a modern world without clocks, watches, or ways other than looking at the sky in order to tell time. Think of a day in your life in such a world. After you spend some time on this, read part *b.*

   b. What did you come up with? What types of individual response, what inefficiencies? Did you find yourself developing an alternate technology to tell time? Did you feel it possible to *have* a "modern" world without time-pieces?

2. Are you an "economic determinist" about technology? Relate your views to the television programs you have seen to date.

3. Do Burke and Cipolla have any fundamental disagreements about the development and role of clocks? Describe the arguments and thinking of each.

PROGRAM SIX

Thunder in the Skies

Program Summary

Our world has immense variety, thanks to the production line. But our life style works only if we have energy, Burke says as he rises over dark Detroit in a glass-walled elevator in the Renaissance Center. Burke then stands by a boiler in the basement of the Center and asks, "What will happen if the cold comes again?" He looks down a tube; at the other end is a fire in the Great Hall of a medieval Saxon manor house 1,000 years ago. A bard sits by it, chanting tales of the outside world to an audience of the lord and lady and all the people of the manor down to the lowest peasant. The fire warms everyone because the life of the community, from eating, to sleeping, to entertainment, is carried on in this great room.

In the twelfth century medieval life changed. The weather grew cold and snow reached far into the growing season. By the thirteenth century Europe was experiencing a mini–Ice Age. Snow and ice made warmth and its conservation more important. The result? It is embodied in great houses like the one we approach with Burke.

Hardwick Hall was built in 1597 yet looks remarkably modern. Burke leads us to the new sources of comfort: a chimney, heating separate rooms; upstairs, rugs on the floor, tapestries on the wall, buttons on the clothes; downstairs, a kitchen with a new, automatic roasting spit; outside, flagstone walks to cover the mud. The division between the social classes had begun, and so had a lot of other things—including a demand for privacy, which could now be realized.

The house also had glass windows—a great many of them. And to make the glass that was now in wide demand, you needed a lot of charcoal, which comes from wood. An ax, biting into the new wood of a standing tree, marks the felling of the great forests of England. The depletion of this resource meant another source of fuel was needed.

Some glassmakers tried moving operations to the colony at Jamestown, Virginia, where sand, needed in glassmaking, and wood were readily available. But that effort failed because of lack of skilled labor and the rigors of survival for the Englishmen in the new land. Breaking glass signals the end of this experiment. Burke sits on a deserted ship as he tells us how the timber crisis was solved by a new way to make glass and work iron, using a new fuel, coal.

The impetus to these new processes was the enormous profit to be made, symbolized by the gold coins Burke sets before us. A metal-making boom increased the use of coal and then, because it was cleaner and made purer iron, coke. Water in coal mines like those stretching out under the ocean off the Cornish coast necessitated an engine to pump out water. Burke stands by Newcomen's great pump that depended on the vacuum formed by condensing steam to draw a piston into a cylinder. With James Watt's improvements and with the introduction of machine-tooled cylinders, the Industrial Revolution (gears, pulleys, engines, and row houses form a montage) was on its way. Life would never be the same.

The next piece of Burke's story takes him to Lake Como in northern Italy. The noxious gases of marshes along the lake's edge gave Volta "inflammable air" in 1776. Burke lights methane above the water and explains how Volta combined it with an electric spark to make an electric pistol. Burke connects the wire and spark; the cork flies past us—the precursor of the spark plug.

The use of electricity in this way worked better with gasoline, the by-product of a fuel used to solve another energy shortage in the mid-nineteenth century. When whale oil for lamps grew scarce, oil from the ground in Pennsylvania was used instead. The next scene is of a gracious German spa where Gottlieb Daimler and his friend found that igniting a mixture of gasoline, made from oil, and air with an electric spark made it possible to build light, movable engines. A succession of horseless carriages provides a background to Burke's explanation of the engine, carburetor, and ignition systems and brings the threads of his story together.

Daimler's engine didn't simply go into cars. We approach an Austrian lake by plane. Here a man named Kress experimented with a new contraption and an engine ordered from Daimler's factory. The experiment didn't work because the wrong engine was sent. Kress was killed. But his idea eventually was taken up by others. Two years later, in 1903, the Wright brothers flew the first airplane.

Program Background

"Thunder in the Skies" is particularly rich in illustrating how the ways in which people lived have changed over the centuries. Manorial life before the twelfth century is recreated at Good Easter, a Saxon manor house in Essex, England. Hardwick Hall—"more glass than wall"—is located in Derbyshire, and the various objects Burke uses to demonstrate the ways in which life had changed all belong to the period.

Halfway through the program, Burke, discussing money, sits aboard a replica of the *Susan Constant,* one of the ships that originally brought colonists to Jamestown. Jamestown itself is rebuilt on the original site of the settlement and is open to the public on a regular basis. Finally, the panorama of the early days of motoring was filmed, most appropriately, on Daimlerstrasse, in Stuttgart, Germany.

Themes

Technology and the way we live. Halfway through the "Connections" series, it may be time to take stock of our technological environment once more. This time let us consider material comfort, resources, and their relationship to population.

Life on the manor shown at the beginning of "Thunder in the Skies" was hardly idyllic. The boys, sitting on the rafter, probably found it quite acceptable, but it is unlikely that we would. The hall was drafty, the fire smoked. The food was, by our standards, plain and uninspired. The bread that formed the mainstay of the diet probably tasted of mold much of the time, or that special mustiness that comes from weevils in the flour. Wearing clothes without buttons is inconvenient at best, and manual labor from dawn to dark to eke out a subsistence living is usually inspiring only to those who don't do it.

"Intellectual comfort" was as hard and plain as physical comfort. There was music, if someone stopped in or if you could make it yourself. The music was limited to voices and a few instruments, and to the tradition that you knew. Books were almost nonexistent, which didn't matter a great deal because few people knew how to read. Contacts with others were similarly few; traveling minstrels like the one in the program must have been a real treat for people who seldom ventured much beyond the manor. Historians have estimated that a peasant in Normandy in the tenth century—not far removed from the time and place of our manor hall—might see only two or three hundred people in his entire life. And that peasant's vocabulary— perhaps 600 words—was as circumscribed as his world.

Life in the towns, of course, was different. You were more likely to die there of disease—but you might have more fun before it happened.

The changes in life style since this time may not all be attributed to technology, but they could not have occurred without technology. And these changes affect more than the immediate comforts of our days. Burke points out that even our physical characteristics have changed. He refers to the mixing of hereditary characteristics that occurred with the advent of railroads in the nineteenth century; most of us have seen furniture and armor from centuries past, all made for people much smaller than we are. We live longer than our ancestors and, perhaps, do not need to be as hardy to survive.

Burke identifies one of the sources of change, however, that has not been affected by technology—it has been, instead, a moving force behind technological development. The weather shapes the way all peoples live, including ourselves. The "mini–ice age" receded in the fifteenth century, but there is a clear possibility that world temperature may drop again. Burke's question, "What will happen if the cold comes again?" is an intellectually chilling one.

Resources and population. None of the changes in people's life styles that we saw in "Thunder in the Skies" occurred without corresponding changes in the environment. The adaptation of resources for our use entails the transformation or destruction of other resources. By the Middle Ages deforestation had become a problem in large parts of Europe; Burke notes this and the shortage of copper for brass during his narrative. Technologically based production, the kind that results in the comforts we enjoy so much, has reached out over the centuries to use more and more of the products of the earth.

The demand for (and the ability to produce) material goods was accompanied by a rise in population. Each fed the other—more food, warmer clothing, and better housing meant that more children survived to become workers and consumers. For centuries this process was slow, but with the Industrial Revolution it, like technology, took off. The first great warning about limited resources—in this case, food—came in the late eighteenth century from the English economist Thomas Malthus. The Malthusian prediction of a population growing until it exceeded the food supply and caused an excess of deaths over births came true again and again, but only on a local scale. In almost all famines, food existed that could have fed the starving people—but it wasn't where they were. Technology has made three things possible and all operate today: first, and most important, is an increase in food production; second is the transportation of food to famine areas; and third, in the long term, people move out of food-poor, crowded areas.

Movements of population, particularly to the New World, brought people in contact with new resources. The result was economic growth of the sort we have seen in past programs, and the creation of systems to make yet more material goods. The questions with which we are left are two: How much do we think we need? How long can we have it?

Readings

Read Chapter Six in James Burke's *Connections.*

A number of articles in the Courses by Newspaper book of readings, *Tech-*

nology and Change, address one aspect or another of resources and technology. Jean Gimpel, in "Environmental Pollution in the Middle Ages" (pp. 135–140), expands Burke's narrative about wood shortages and goes on to discuss the air pollution that resulted from the use of coal. Similar shortages now face the United States, but in mineral resources rather than in wood. Walter R. Hibbard, Jr., in "Mineral Resources, Challenge or Threat?" (pp. 130–134), suggests ways technology can be used to conserve and expand these resources, rather than squander them. René Dubos, in "The New Environmental Attitude" (pp. 141–144), examines what he believes is a new attitude—"the willingness to abandon policies and practices that are technologically possible and economically profitable but socially objectionable." The Malthusian dilemma is re-examined in its modern context by Nathan Rosenberg in "Economic Growth, Technology, and Society" (pp. 222–229), and Carter Henderson describes one type of response to fears about resource depletion in "The Frugality Phenomenon" (pp. 230–238).

Program Six begins with the problems of producing food under changing—and difficult—conditions. The relationship of a population to its food supply has always been a critical variable in a society. Neal F. Jensen presents a disturbing picture in "The Food-People Problem" (pp. 186–192), in which he argues that we already put more nonrenewable resources into food production than we receive from it and still the world's population isn't fed adequately. He also points out that we may soon reach limits in increasing yields per acre. "The Migrations of Human Populations" (pp. 175–185) by Kingsley Davis adds to the gloom; Davis discusses the functions of migration in the past and present, but points out that the world is now full.

Kingsley Davis raises another disturbing aspect of modern technology in relation to food in the seventh newspaper article, "Technology, Population, and Resources." In the United States we now "spend" more energy to raise food than we get back from it. Clarence J. Glacken, in the fifth newspaper article, which deals with the problem of resources, looks at some of the ideas we have held concerning our relationship to the environment. He suggests in "Culture: The Link between Nature and Technology," that if we look at the history of the way man has acted on the environment we will have a different picture than if we consider the history of technology.

Discussion Questions

1. At the beginning of "Thunder in the Skies" Burke asks, "What will happen if the cold comes again?" If the average temperature of your area were to drop ten degrees, what would the impact be on your life style?

2. Imagine yourself living in either Hardwick Hall or the Saxon manor house. How would that life compare with your own?

3. Kingsley Davis raises a number of questions about population and food in his selection in the book of readings and in his newspaper article. Identify the three you consider most important. How satisfying are the possible solutions to your questions posed by Rosenberg and Henderson?

4. How many of the sources of energy and of pollution identified in "Thunder in the Skies" and in the readings are still with us today? Do we know of solutions for the types of pollution these energy sources cause?

5. Pick a meal—breakfast, lunch, or dinner—and list the points at which energy was used in producing the food, getting it to you, and preparing it.

The Long Chain

Program Summary

James Burke greets us from what appears to be a giant metal box. This box is a part of a method of transportation whose impact today is comparable to that of the fluyt, a Dutch ship in the seventeenth century. Today, the Boeing 747 air freighter brings change in its wake; so did the fluyt. Indeed, the fluyt set in motion a whirlwind of events that shaped the known world.

To demonstrate the construction of the fluyt, designed in Holland, a succession of graphics strips the common fighting ship of the day to its hold, alters its shape, and builds it up to carry cargo. The last drawing dissolves into a model of the little fluyt that made the Dutch rich.

The Dutch, however, weren't the only trading nation. The English also wanted to play the trading game, and it wasn't long before they entered it successfully. Seventeenth-century traders discuss business over coffee as Burke describes the "triangle trade"—cotton, slaves, sugar—that was carried by ships insured with Lloyd's of London. Lloyd's identified the ship's hull as a key factor in safety, and it is here that we see the first link in the chain that came from coal tar.

Pitch, brought first from Finland, then from the American colonies, was needed to protect ships' hulls from destruction by a tiny worm that lived in tropical waters. The American Revolution, however, cut supplies. A fife and drum signal a change in scene to the Scottish village where a down-at-the-heels "laird," the Earl of Dundonald, went broke trying to extract pitch from coal. But as was his luck, he succeeded in developing the process just as the English navy began to plate ships' hulls with copper, making pitch obsolete for the purpose. He died in poverty without benefiting from coal tar or from a by-product of coal he had accidentally discovered—gas.

In the early nineteenth century, gas provided lights for the city of London and for the stage on which Burke stands to recite the wonders of this modern invention. The "modern invention" was widely adopted. It made possible a whole new way of life, with evening activities from school to theater; but gasmaking also produced another by-product, coal tar. At first coal tar was dumped into the Thames and other rivers, but ultimately it spawned many products.

The first was the mackintosh, a waterproof coat made of rubber dissolved with the coal-tar derivative naphtha. The manufacturers of mackintoshes tried to get British botanists to grow rubber in India so it wouldn't have to be imported from South America. The botanists, however, were more interested in growing cinchona in India. From it quinine was extracted to protect British colonists, who were hacking plantations out of Asian jungles, from malaria. That experiment failed, but an attempt in 1856 to manufacture quinine from coal tar resulted, quite accidentally, in the first aniline dye.

As a result of aniline dye, Victorian England burst into color, and its inventor, William Perkin, grew rich. The British, however, still regarded manufacture as a lower-class activity, and they blew their lead. The double-eagle and brass bands of the Germans replace the Union Jack. To boot, the Germans developed a lot of other products from coal tar.

Burke, driving an old combine across a wheat field, stops to explain a peculiar economic situation of the late nineteenth century. The McCormick harvester was largely responsible for producing great quantities of American wheat, which forced the world price of wheat down. The German government subsidized the export of Prussian rye to compete with the wheat. But the Prussian peasants couldn't eat their own rye because it cost too much. Burke throws a loaf of rye over one shoulder. The peasants couldn't eat American wheat either, because German import duties made it cost too much. Burke throws a loaf of wheat bread over the other shoulder, demonstrating the plight of the starving peasants.

In order to boost its own wheat growing, the Germans needed cheap fertilizers. The way they solved the problem is a bit of a chemistry lesson. Burke, in a laboratory, explains the Haber Bosch process to produce ammonia for fertilizer. Then, with old movies of Paris in the background, he explains how this discovery was made useless—for the time being—by the French discovery of calcium carbide. The Germans combined the calcium carbide with nitrogen for use as fertilizer, and its by-product, acetylene, was (less successfully) used instead of gas lamps. Germany, with cheap fertilizer available, was off to becoming a major industrial nation. Partly because of the rivalries this caused, Germany also became a participant in World War I, when the Haber Bosch process became useful again. This time it was used to make explosives.

The one thread left hanging is what happened to the acetylene. Color movies of the 1939 World's Fair in New York and the DuPont pavilion introduce us to the tapestry it wove.

The thread is nylon.

Program Background

"The Long Chain" presents a fascinating succession of vehicles. It begins in the hold of a Boeing 747 freighter at the Frankfurt airport in Germany. Almost immediately it moves to Hoorn Port in Holland and a graphic analysis of the fluyt. The train on which Burke travels in Malaysia (where he also drinks the gin and tonic) is typical of those on the Malaysian Rail system, where prices for a stately (if sometimes hot and sooty) journey are low. The combine that Burke drives is a working model, but the flags that make it ominous during the first seconds of that scene were put on it for this program.

The suspended monorail Burke travels near the end of the program is in Wuppertal, Germany. Although it looks futuristic, it was opened by the Kaiser in 1900; it is still in operation today.

Themes

The impetus for the development of technology can come, as we have noted before, from many sources. "The Long Chain" illustrates a number of these, but two are particularly outstanding—science and engineering and the way they are brought together in a process of discovery.

Science, engineering, and technology. How do we distinguish among science, engineering, and technology? You can probably formulate your own definitions for each. Others have formulated theirs, and there are many disagreements about the boundaries. For our purposes, a broad picture of the three will do, and we can start our sketch by penciling in the outlines of the discovery and manufacture of aniline dyes. They were discovered, as you will remember, in England by William Perkin while he searched for synthetic quinine. Perkin became rich, but England—where inventors were held in relatively low esteem and where investors could find other, surer, outlets for their capital—took little advantage of its lead. Further, Perkin retired to a "life of science" at an early age—exactly the kind of thing a University man would want to do in that era.

Germany, on the other hand, had *technical* schools which cooperated with universities and industry. Their graduates, who did not scorn applied knowledge, took coal tar and aniline dyes and developed a wide range of products.

This difference finds a parallel in science and technology. Very generally, science is systemized knowledge. "Science" implies the exploration and organization of knowledge for its own sake. Technology, on the other hand, is the application of knowledge, whether based on science or not, for practical purposes. Engineering, in its broadest sense, is the process through which this is done. The English, if we may return to our example, honored scientists; the Germans rewarded engineers.

The distinction, however, is not absolute. Discoveries, particularly of the kinds that are chronicled in "The Long Chain," can be both "science" and "technology" at once. The discovery of nylon, for instance, contributed to our knowledge of molecular structure and bonding while it produced a new product that we have used in many ways, from clothes to tires.

The further back in time we go, the easier it is to see the distinction between science and technology. The fluyt was a technological innovation; its building neither sprang from nor resulted in new scientific knowledge. The botanists who discovered the cinchona in South America were cataloguing plants in the interests of science. Usefulness was a by-product. The botanists were not blind to the possible existence of such by-products as quinine, but they were not searching specifically for a cure for malaria.

The role of science and engineering in promoting technology is equally complex. The engineering skills that produced the fluyt also provided voyagers with an ability to explore the New World. Engineering provided gaslight to the city of London. The scientific inquiry that discovered the rubber plant eventually led to the development of a technology for its use.

Perhaps most importantly, the mode of inquiry developed for science contributed to both scientific and technological discovery. The systematic exploration of reasonable hypotheses, the drawing of conclusions concerning causal events and plausible outcomes, and the method of logic upon which these depend are the basis for the kinds of developments Burke chronicles in this and other programs.

Readings

Chapter Seven in James Burke's book, *Connections,* expands the information presented in this program.

The relationship among science, engineering, and technology is the theme of several articles in the Reader, *Technology and Change,* and in the newspaper series. Both Robert P. Multhauf, author of the ninth newspaper article, "Science and Technology," and Peter Drucker, in his Reader selection "Applied Science and Technology" (pp. 245–250), believe that science and technology have come together in the twentieth century. Multhauf argues that there was an earlier period when the "scientist-inventor" existed, but that the distinction between science and technology took hold again by the eighteenth century. Both, interestingly enough, identify Perkin's discovery of the first aniline dye as the point at which the two again begin to merge.

In his article, Drucker distinguishes between the "technology" practiced today and that of the last century. The differences center on the systematic way research is now conducted, as opposed to the more hit-and-miss methods of earlier days. Charles Weiner's account in the Reader of the emergence of the transistor (pp. 251–261) illustrates this point.

The "density" of innovation depicted in "The Long Chain"—the number of discoveries occurring at a particular point in time—parallels the growth of science noted by Derek de Solla Price in his selection, "Little Science, Big Science" (pp. 239–244). As we have seen, the television program dwells on each innovation for some time at the beginning, then picks up speed with Perkin's discovery in 1856. The pace, like the exponential growth of scientific knowledge in the twentieth century, increases until we are presented with the wealth of products displayed at the New York World's Fair.

If the goals of the scientist are to extend the boundaries of knowledge, the goals of the engineer are more earthbound. Eugene S. Ferguson outlines some of these in the tenth newspaper article, "The Imperatives of Engineering." He discusses several characteristics of an engineer's approach to problems: a concern for efficiency, a preference for labor-saving systems, a desire to wield control, an urge to transcend human scale, and an involvement with a problem or project. Each contributes to the engineer's ability to solve technical problems, but each can also be the source of unforeseen complications. This article is also pertinent to the themes of work and the social context of technology in Programs Two and Four and will be raised again in connection with Program Nine.

Sir William Fairbairn, in his Reader selection "The Engineering Profession" (pp. 267–272), presents a picture of the engineer that contains fewer qualifications. Many of the engineering accomplishments he traces through history have been shown on the programs in the "Connections" series. Fairbairn's use of the term "engineer" is broadened by Thorstein Veblen in "The Role of the Engineers" (pp. 283–289) in order to apply "engineering" to the entire production process rather than just the mechanical aspects of it.

Finally, another aspect of the relationship between science and technology is discussed by Jerome R. Ravetz in "Social Problems of Industrialized Science" (pp. 262–266). Ravetz believes there are problems that arise specifically from the modern merging of the two. "Industrialized science" now deals with a technology that offers bizarre possibilities but operates without the constraints imposed in traditional science by a group of peers, and it is no longer judged by absolute utility or profitability of a product as, in the technology we have seen in "The Long Chain," gas for lamps or the uses of coal tar.

Discussion Questions

1. Referring to the first seven programs in the "Connections" series, identify at least three instances each of (1) scientific discoveries resulting in tech-

nological change; (2) engineering being responsible for technological change.

2. Both Ferguson and Ravetz are concerned about undesirable outcomes of science, engineering, and technology. Using their specific concerns, identify at least three situations in "The Long Chain" for which you could project unfortunate effects.

3. Compare Veblen's picture of the engineer with any two of the inventors featured in "The Long Chain."

PROGRAM EIGHT

Eat, Drink, and Be Merry

Program Summary

The objects around James Burke in a lighted office are plastic and, like the modern world, change their shapes regularly. A dizzying succession of machinery, illustrating change, takes us to a factory that helps finance our technology-rich life style. It makes credit cards, an artifact that may signal the logical outcome of a process of credit and borrowing that began about 600 years ago. The factory door slams across the screen and we are taken to the court of the Dukes of Burgundy in fifteenth-century France.

The fourth Duke of Burgundy waged war on credit. He lost both his battles and his money, and his defeat was the result of a change in the face of war. Burke, in a war games room, explains the successive use of pikes in 1476, the arquebus in 1503, and the flintlock in 1521. The tactics that accompanied these changes were also altered, and each change involved more and more soldiers.

More and more soldiers required more and more food as they marched to battle. The matter reached a crisis stage for Napoleon in 1800 in Marengo, Italy, where he almost lost a battle because his men were out looking for something to eat. A column arrived in the nick of time and Napoleon, like Burke, was able to dine on his chef's latest concoction—Chicken Marengo.

When he returned to France, Napoleon, in an effort to boost the French economy, called for profit-making inventions, including something to feed his armies. Burke leads us to the invention that solved Napoleon's problem through the vineyards of Champagne. But instead of the pale gold bubbly we expect from the champagne bottles, peas, beans, and soup tumble out. Nicholas Appert had discovered that food could be preserved by heating it in sealed bottles.

46

The change from bottles to tin cans for food preservation followed a strange series of links. The end of the paper chain Burke uses to demonstrate them was a man who bought a translation of Appert's book and who, because of his printing background, knew metal technology. Even the Queen, tasting his new canned food in 1813, pronounced it "Delicious!"

But there were problems in canning. During the Crimean War in the 1850s, many cans of meat spoiled. The reason mistakenly attributed to the spoilage was heat and bad air, and these also concerned a man in Florida. Heat, stagnant water, and bad air—"as everyone knew"—caused both food to spoil and people to get fever, and there was plenty of all three in the swamps of the southern United States.

John Gorrie, the mayor of Apalachicola, thought he could cure people of malaria if he could keep them cool. He had some success with his technique because it involved putting gauze over windows which kept out disease-carrying mosquitoes. But he had a bigger success with his machinery to keep the air cool. What he had developed was an air conditioner which, Burke tells us as he whips out a glass, also made ice.

The same refrigeration system was used to bring chilled beef to a banquet in New Orleans in 1850 and frozen beef to the docks of London from Australia in 1880. The next step, however, was linked to the beer-drinking habits of the Germans. Scenes of the Munich festival and great steins of beer provide a background to Burke's discourse on the Germans' need to cool beer vats that would permit brewing all year round.

They solved their problem by adapting Gorrie's system to use ammonia rather than air. This is the system our refrigerators use today, but we have increasingly used gases that evaporate at lower temperatures or higher pressures. Such gases are hard to keep in liquid form without a safe container. The answer was the vacuum flask, invented by Violle in 1882 and refined by Dewar in 1890. It is introduced to us at an Edwardian picnic and brought to a modern concrete landscape. The greatest impact of the vacuum or "thermos" flask on the twentieth century was made by the rocket moving past Burke. If you take a vacuum flask, remove the top, release a combination of hydrogen and oxygen into a confined space with a small hole in it, then ignite them, you have a rocket that will take off.

Program Background

Food is a major theme in "Eat, Drink, and Be Merry," and a number of the major scenes are related to it. Chicken Marengo is served to Burke in a farmyard on the edge of the battlefield in Italy for which it was named. Peas, beans, and soup are served from champagne bottles in the Petit Trianon, the little palace built for the Napoleonic court, in Epernay, France. The French

Consul's dinner, where the wine was chilled, was reconstructed in a period house in Apalachicola, Florida, the home of John Gorrie. The sickroom episode was filmed in the Gorrie museum there, as was the ice-making unit. Both the sequences about chilled or frozen beef were also filmed on location, in New Orleans, Louisiana, and at the Smithfield Market, where the first Australian beef arrived.

Burke demonstrates the operation of an original papermaking machine in the Museum of Papermaking, Grenoble, France. And Burke was on the scene to film the Voyager lift-off of August 23, 1977, especially for the closing sequence of this program.

Themes

War as impetus to technology. Conflict seems, at times, to be as much a part of human existence as food. At other times, it seems to be even more important. In the area of technological growth, armed conflict has certainly had a more consistent impact on change than the pursuit of more and better food. Even "Eat, Drink, and Be Merry" has, at its heart, military need rather than a desire to provide people with a better life. Why is this?

One reason is that the need to fight "better" is very immediate. The rewards are clear cut. So are the penalties for failure. Innovations can be tested quickly and with obvious results. The knights on horseback in Program Two had a real incentive to make better weapons and breed stronger horses— self-preservation. Similarly, governments find investing in research for warfare both useful and rewarding—and this investment extends far beyond the implements for battle. "An army," Napoleon is supposed to have said, "travels on its stomach." He, if you will remember, provided the impetus for the development of preserved and canned foods.

Another reason may be that war increases mobility and, through it, the "spread" of new ideas. As the armies Burke describes grew, more and more men (not to mention camp followers) were brought from towns and villages into great concentrations, then marched to foreign climes. They often caught a number of exotic diseases from one another in the process, but, more importantly, there was also a gathering of individuals with different skills and traditions who had a common purpose. One result was an accelerated diffusion and adoption of new ideas.

Government and technology. "Eat, Drink, and Be Merry" also demonstrates the relationship of government to technology. Napoleon's government offered prizes for ideas that would help French industry; Gorrie obtained a federal patent for his ice-making machine. Further, governments subsidize research, they enable individuals to profit from discoveries by protecting their rights, and they regulate technology. Government intervention can either be helpful to technological advancement or it can impede it by (1) withholding incentives for it; (2) banning technological research; or (3) banning the application of technological discoveries.

Governments also provide a social, legal, and economic infrastructure within which technological development takes place. Governments regulate systems of ownership and taxation. They provide incentives or disincentives to trade and commerce. They establish and maintain standards for weights and measures. They are often responsible for communications systems— mails, telecommunications, transportation—that facilitate the movement of knowledge and goods.

Finally, to close a circle we started at the beginning of this section, governments make war.

Readings

The developments in this program are discussed in Chapter Eight of *Connections*.

Several articles in the *Technology and Change* Reader address war and technology. Quincy Wright traces the history of war in his Reader article, "Technology and Warfare" (pp. 302–311). His article should help you place into a more coherent perspective the material you have encountered in this program and Program Two, and that you will meet in the next "Connections" program. Like James Burke, Wright sees the kinds of technology affected by war increasing over time, not simply because of a growth in technology but because of a growth in the areas of life affected by a nation at war. Giulio Douhet foresaw this line of reasoning in his selection "The Command of the Air" (pp. 312–316). He believed the airplane would do more to involve the civilian population of a nation in battle than any other military invention. Lynn White, jr., in his article "The Act of Invention" (pp. 379–391), adds material on several of the types of warfare we have seen.

The development of bigger and better ways to kill and maim human beings has always raised moral questions. These took a quantum leap with the first atomic bomb. Three selections in the Reader, by Albert Einstein, the General Advisory Committee to the Atomic Energy Commission, and Andrei D. Sakharov (pp. 317–331), present some of the doubts that scientists have concerning nuclear weapons.

Herbert F. York and G. Allen Greb relate the technology of war and the role of government in "Wars: Hot and Cold," the eleventh newspaper article in the Courses by Newspaper series. In it they describe the ways in which modern governments influence military research and development. A. Hunter Dupree, in the twelfth newspaper article, "The Government's Role in Technological Change," explains some other ways government can influence technology.

Three articles in the book of readings describe the direct intervention of government into the development of technology. Both Reynold M. Wik (pp. 338–347) and Carroll W. Pursell, Jr., (pp. 348–355) describe how the government supports technological research, Wik in agriculture and Pursell in industry. John G. Burke's selection, "Bursting Boilers and the Federal Power" (pp. 356–367), concerns government regulation and its consequences in a particular case. The effects of this intervention on technology itself are explored in the Reader articles by Sanford Lakoff (pp. 368–371) and Jerome R. Ravetz (pp. 262–266), which were described in Chapters Two and Seven of this Guide.

The possible consequences on society of government intervention are described in two Reader articles: David E. Lilienthal is eloquent about the benefits of the TVA in "Democracy at the Grass Roots" (pp. 428–431); Arthur E. Morgan is equally eloquent about a less appealing aspect of dam building in "The Garrison Dam Disaster" (pp. 432–440).

Discussion Questions

1. What different types of technology shown in "Eat, Drink, and Be Merry" did warfare affect?

2. Compare James Burke's and Quincy Wright's methods of relating military history to technology. Do they differ in their interpretations of events? You might want to review your conclusions after you have seen Program Nine, "Countdown."

3. Do the descriptions of government intervention in the readings discussed above have parallels in the television program? In past television programs? What means of government intervention in technology are apparent to you in your immediate environment?

4. Basing your answer on both the program and the reading material, explain whether you believe the morality of warfare has changed over the past thousand years.

PROGRAM NINE

Countdown

Program Summary

The Apollo moon shot is a perfect example of why technology is a double-edged weapon. The same instruments that control the great rocket can make plane travel safer or steer atomic missiles. Their effect on military strategy may be as decisive in altering the course of history as another missile, the cannonball, was 600 years ago.

Fires and explosions fill the screen. A moving cart transports us to the walled city of Aigues-Mortes, France, a perfect example of fourteenth-century pre-cannon fortification. Burke explains the vulnerability of the curtain walls and corner towers to artillery, then takes us to another wall. At first glance it looks like a park, but the grassy mounds actually belong to a "star fort." Lines of fire superimposed on an aerial view of the town illustrate its defensive power. Burke, now standing opposite the walls, continues with a description of the calculations needed to shoot cannon accurately at the city and demonstrates the instruments used to make those calculations.

The same calculations helped surveyors (which aided the military, who usually wanted to know where they were) in their work for Henry VIII of England. Henry dissolved the monasteries in 1536 and confiscated and sold their land. The buyers, quite naturally, wanted to know exactly what they were getting so the land was surveyed. These surveys also laid the groundwork for the first mapping of England.

England was mapped by the late sixteenth century, but it was not until 1825 that surveys of the entire British Isles were completed. Burke appears climbing into sight as he explains the role of Wade's road in bringing the surrounding Scottish Highlands under London's rule in the mid-eighteenth century and, eventually, in creating a map of the area. Next, extending the surveying to map Ireland resulted in a new kind of illumination, limelight. The limelight was bright enough to provide a siting on mountain tops so the distances among them could be calculated.

The bright limelight was also tried in lighthouses but proved too expensive. Burke appears on stage, "in the limelight," to dramatize the new solution.

With pictures, curtains, and flourishes, he explains the events that led to the dynamo, which powered the arc light. The arc light, illustrated in the sequence that begins each "Connections" program, works because electricity jumps from one carbon rod to another. The world's first electric light had come into being, as well as the first part of an invention that is now in your house.

The next part is introduced by a disastrous explosion in England and continued in a Colorado saloon. "Gun cotton"—cotton treated with acid—and billiard balls were, indeed, an explosive combination.

The third piece was as showy but quieter. The most striking component was Leland Stanford's bet about the way horses ran. From a series of photographs the experimenters learned that all four feet leave the ground as a horse gallops. More importantly, they also found that if the pictures were run rapidly through a projector, it looked like the horse was really moving on film.

The last part of our modern invention was the telegraph, used initially to prevent trains from running into one another by signaling their location. And Thomas Alva Edison put the pieces all together.

He used the light bulb he had invented (on the principle of the arc light) to illuminate sequential pictures on a celluloid strip made from, among other things, the outer materials of the explosive billiard balls. "Old time" movies of earlier program sequences run by quickly as Burke recaps the processes that led to this point. Sound was added from the recording device Edison had made (derived in principle from the Morse telegraph) and used to record "Mary Had a Little Lamb." A few years later pictures and sound were put on film, then transmitted electronically. Burke faces us and taps the back of the screen saying, "This is it." "This" is your television set.

But we haven't quite reached the end. Television isn't an end-point. It is part of an ongoing process. It acts as a relentless reminder of the pace of contemporary change. It also acts to accelerate change. What will it, and the kind of science that invented it, do to change the future? What will it do to us? These are some of the questions that will be taken up in the next, and last, program.

Program Background

History buffs may discover two new places to visit during "Countdown." The first is Aigues-Mortes, on the Mediterranean coast in southern France, one of the few remaining untouched examples of pre-cannonball fortification. It was completed in 1300, several centuries before the next site was built. Maarden, Holland, is a star fort that has survived civic expansion and modernization. The aerial shots provide the most complete illustration of the pattern and planning imposed by the cannon.

On the other hand, the ability of the camera to create the illusion of reality where none exists is beautifully demonstrated in the scenes of Little Jack Horner's Pie. The Life of the Glastonbury Abbey Kitchen appears real, but when "Jack Horner" disappears against the ruins of the modern abbey, it becomes clear that the camera initially excluded the signs of decay.

Also created—in the BBC studios—were the theater set and the Old West saloon. The gun-cotton explosion, however, was reconstructed at the original site in Faversham, Kent, England, and Edison's Menlo Park Laboratory is now part of the Greenfield Museum in Dearborn, Michigan.

Themes

James Burke has pointed out that technological innovation is the result of many factors. "Great discoveries" are not the product of lonely, uninformed labor in a garret. And yet, many of the innovations we have seen are the work of individual men or women. "Countdown" and the other programs in the "Connections" series are full of the names of the people who put bits and pieces together and came up with something that no one else had done before. Why them and not someone else?

One of the answers that Burke has suggested is that these people were at the right place at the right time. Thomas Edison, for instance, worked at a time when developments in a number of fields occurred and could occur, in a place where he could get information about them, and in a situation where he would be rewarded for pushing these developments a few steps further.

But Edison wasn't the only person who lived in the right place at the right time and wanted to do things. Neither was Thomas Drummond, the inventor of the limelight. Why *these* two and not two others? Is it only because they, like Alexander Graham Bell, were "first to the patent office"? It is true that many scientific and technological advances are made simultaneously in a number of places, but time and circumstance still aren't enough to explain the enormous productivity of some individuals and some cultures. Three other factors, applicable to both individuals and their societies and interactive to a great degree, are also important in the inventive process.

Individual creativity. There is an old game in which you are handed six matches and asked to make four contiguous triangles. After various attempts at putting them end to end or across one another somebody puts three on the table in a triangle, then builds a three-sided pyramid above them with the remaining three. That individual either knew the answer or approached it creatively.

Creativity involves "divergent thinking"—or thinking that deviates from customary patterns. The creative person looks for solutions outside the patterns and formats that most of us see. Although there have been a number of theories that attempt to explain creativity, none of them is completely satisfactory. All we know is that it exists, and some people seem to have more of it than others.

Training. The contribution of education to technological innovation, like that of creativity, is hard to pinpoint, but it exists just as surely. Education need not be formal, but an individual must have learned something, somewhere, before he or she can change or improve it. Program Eight, "The Long Chain," attributed Germany's lead in chemical technology to technical schools and their links with industry; none of the people mentioned in "Countdown" was working in an unknown field—all had had some training. And the more people who have a particular type of knowledge, the greater the chances that advances will occur.

Further, none of the individual inventors we have seen was illiterate. In addition to the training necessary for invention in a given area, the ability to read and write is of tremendous importance in the history of technology. We have seen the effect letter writing, for example, has had in several programs—the sequence of letters that enabled Pascal to invent the barometer, the communication that led to preserving food in tin cans, and Priestley's letter to Alessandro Volta about his soda water and work on gases. Literacy—taken for granted by most of us—allows people to read of what others have done, and to record their own activities, both to remember them and to provide their knowledge to yet others.

Intellectual freedom and economic mobility. Finally, individuals need to have some prospect of gain from invention. Both intellectual and economic mobility are important in this regard: Intellectual freedom provides a flow of information and encourages people to develop their abilities; economic mobility provides an incentive to develop new ideas.

Readings

Read Chapter Nine in James Burke's book *Connections.*

It is virtually impossible to address the question of impetus to invention within individuals without considering a host of other questions. The selections in the Reader, *Technology and Change,* and the newspaper articles related to this program raise many of them as they discuss the individual and inventiveness.

Jewkes, Sawers, and Stillerman in "The Sources of Invention" (pp. 406–416) start by asking, "What is an invention?" The answer is as difficult to isolate as "who" the "inventor" is. The selection ends with another broad query—whether our society is set up in such a way as to encourage that inventiveness.

Interestingly enough, the example used here to illustrate the complexity of "invention" is nylon; you might compare their treatment of the topic to Burke's in "The Long Chain."

Two Reader selections and the tenth newspaper article present different aspects of the thought processes of individuals who deal with technology. Eugene S. Ferguson's newspaper article, "The Imperatives of Engineering," was recommended as background reading to Program Eight; it is also applicable here. The same author raises a different aspect of how engineers think in his Reader selection, "Nonverbal Thought in Technology" (pp. 400–405). Ferguson argues that much technology is visually and artistically oriented and that we are doing ourselves a disservice by insisting that engineers be trained to apply rigorous mathematical proofs to everything. "Freedom and Direction" by James C. Wallace (pp. 441–444) also raises the unique approach of engineers to problems and tasks, and questions the ways in which the "engineering ethic" is allowed free rein to explore and develop technology.

Burke's questions in the last part of the program are taken up and explored by Kenneth Keniston in "Technology and Human Nature" (pp. 445–448). Not only do individual make-up and intellect determine the course of many inventions; technology, Keniston argues, also may be reshaping human nature and, if this is so, we must be alert to the ways in which this might express itself.

Lynn White, jr., in the thirteenth newspaper article, "The Mystery of Inventiveness," illustrates some different approaches to technology that are further discussed in his Reader article, "The Act of Invention" (pp. 379–392). Jacob Schmookler (pp. 396–399), cited in Chapter Five, also presents some aspects of the psychology of invention.

Discussion Questions

1. By this program, you should recognize subjects and inventions that have appeared before. Choose one of the following and integrate its treatment in "Countdown" with related material in previous programs: Illumination; War; Communication.

2. Outline the qualities you believe Burke ascribes to inventors, then check them against the material in the "Connections" programs or book. How do they compare with the articles by Ferguson and Wallace? Are there differences between the portrayal of scientists and engineers in the television programs and in the readings?

3. Have you read a biography of an inventor? What does it say about the sources of inventiveness for that individual? How does this compare with the analyses in the "Connections" series and the readings?

PROGRAM TEN

Yesterday, Tomorrow, And You

Program Summary

"Yesterday, Tomorrow, and You," the last program in the "Connections" series, summarizes the themes that have been presented thus far and projects them into the future. It opens with a discussion of the changes of nature and the great leap in the pace and kind of change when humans began to intervene in shaping their environment. It continues as Burke names the eight major inventions whose history he has highlighted, and examines their place in our interdependent, changing world. Scenes from the first program illustrate the ways in which technological change can act as a trigger for other, multitudinous inventions.

Today, because of the exponential growth of science and technology, there are thousands of potential triggers of change. Burke, starting with the technology available in a modern hardward store and moving into a classroom, discusses some of the linear views of change we have traditionally been taught. These approaches, he says, neither explain nor predict. Change and its effects are too complex. Recapping sequences from previous programs, Burke reviews our exploration of this complexity in the past. Will understanding those sequences enable us to predict change in the future?

Not completely. There are too many factors and too many things we don't know. Burke holds up Volta's electric pistol. What would you have made of this? As with a jigsaw puzzle, if you don't have all the pieces, you can't see the whole picture. He gets in a car and drives off.

The car is a vehicle for Burke's summary of communication and change in history. As knowledge grew, so did the kind of specialization necessary to produce the objects we use. And many technologies are so far beyond our comprehension that we are reduced to judging them by our gut reactions.

Burke drives into Cape Canaveral and asks, "How can you evaluate research such as this if you base your evaluation on emotions?" But most of us have no other choice. The frequent result—in this case, a broken tangle of unused space equipment—is a tremendous waste both in the present and for the possible future.

The nuclear power plant Burke visits next represents some of the dilemmas of modern technology. What do we do when we want the things (in this case, power) technology provides, yet are uneasy about possible side effects? Burke walks over a tank of water holding radioactive fuel and begins to offer possible answers. None is completely acceptable.

We must know what we are doing before we can plan for the technological future, but how can we ever learn enough to participate in the decisions that will be made? Burke strides among computers and poses the dilemma of the average person. It looks as if there may be only two options. (1) Do nothing. (2) Strike out at the things you don't understand. The computer center dissolves into patterns of violent destruction of technology in slow-motion sequence.

The program ends with Burke talking quietly. These actions aren't solutions either, he says. The key to change is the key of knowledge. Today, scientists and technologists have that key more than anyone. But all of us can learn from one another because none of us knows everything. What we must do is try to understand, and ask questions about those things we do not understand.

Where do we start?

Ask yourself, Burke suggests, if there is anything in your life you want changed.

The faces we have seen in previous programs flash by us, interspersed with scenes from a roller coaster. The series is over.

Program Background

Cinematic technique dominates the last program in the "Connections" series. The visual effects of the "smashing modern technology" sequence is an arresting feature of this program with its combination of quick cuts and slow motion. Another shot in which the program's narrative is enhanced by visual effects is the movement toward Burke when he is seated in the British Airways Computer Complex at Heathrow Airport. The alternation of one computer after another provides a counterpoint to Burke's discourse as the camera rolls toward him. And there will be few viewers whose stomachs didn't lurch as they watched the roller coaster scenes that close the program.

Some of the techniques are, as in other programs, enhanced by on-location filming. The nuclear power station is on the Severn River, and the medieval water mill is in Mapledurham, both in England. The colloid chemistry sequence was filmed at the United Kingdom Atomic Research Laboratory at Harwell, England, and the eight artifacts whose histories we have seen were assembled in a B-52 at Sawyer Air Force Base in Marquette, Michigan.

Themes

At the beginning of this Viewer's Guide we raised certain broad themes to provide a framework for subsequent questions and discussions. We are now left standing squarely between the past—some of which Burke has shown us—and the future. It is time to re-examine these themes with reference to the future.

The result of our examination might best be summarized as a series of statements. Many are simply illustrations of how little we know.

1. There is no single "source" for technological change. Each innovation and its results spring from a web of interrelated factors that may or may not include the following: time and circumstance; need; economic factors; social and cultural influences; war; government; religion; communication; individual creativity; the general level of education and, by extension, the "pool" of available minds to confront a problem; science and the organization of research; pure accident.

2. Technology affects us as individuals and as members of a society. We affect technology. We can identify some clear relationships, but are not sure of all the ramifications of influence in either direction.

Our lives are profoundly influenced on the most material everyday level by the artifacts we use and the food we eat. Medical technology has saved many of us from early deaths and enabled most of us to interact more fully with the world around us by correcting physical disabilities. The imperatives of time and the way in which we organize our work determine the rhythm of our lives—for some of us these imperatives are far more influential than the ancient succession of weather and seasons. Each of us as an individual faces less mortal risk in living (that is, our chances of living longer are better) because of technology, but the fabric of our society may be at greater risk. A political upheaval in the Middle East that closes oil production, a crop failure in Canada, or a guerrilla group with an atomic bomb can cause ripples of effects that spread throughout the world. Technology also affects our political behavior. The need to play to the media, according to many critics, has distorted issues in national political campaigns. In addition, we are all affected by our natural environment and the ways in which technology shapes it.

3. We are not sure whether we need to "control technology," how to do it if we want to, or what effects such control would have. The "systemic nature"

of technology means that if you tinker with one aspect, all kinds of results follow, many of them unexpected. We don't know exactly why inventions occur, what they will do, or what changing some part of the system will do. This complexity is increased because there are too many branches of specialized knowledge for one person to master them all, or even be able to talk to practitioners in a majority of them. And in the modern world, everything happens quickly. Innovations spread and their consequences are upon us before we even know what they are.

But all the "don't knows" do not consequently mean we are helpless. We can act to question and attempt to control technology, but we need to act with the imperfect tools at our command and to proceed cautiously with a full realization of our ignorance. James Burke simply asks us to recognize the complexity of our world and the diversity of the activities it harbors. There are no simple answers; there are very few simple questions.

Readings

Many of the complex issues raised in the television program are recapped in Chapter Ten of James Burke's book, *Connections.*

James Burke has left us in the television program with the questions that began the Courses by Newspaper readings. Most of the articles in the first section in the Reader address the role of technology in our society and our ability to control it. Several were recommended to accompany the first program; they could well be reviewed at this point.

Another article suggested for an earlier program, "What Computers Mean for Man and Society" by Herbert A. Simon (pp. 68–76), is worth reviewing for its qualifications of James Burke's statements on computers. It is true, as Burke maintains, that computers enable us to predict what will happen—but only if we know all the factors that will affect an event. Throughout the programs in this series, however, we have seen that the sources of change in any segment of society are complex and difficult to identify in exact proportion and detail.

A further area of implicit dispute concerns the future. Burke outlines several alternate scenarios. We can imagine a number of authors of both newspaper articles and Reader selections arguing that these are incomplete. You have already read some of these, but one that has not yet been suggested is the view of technology presented by Clarence E. Ayres in "The Industrial Way of Life" (pp. 425–427).

One of the questions raised repeatedly in the Connections material is whether "big is bad" in technology. Wilson Clark describes a creative response currently being advocated in many developing countries. This is "Intermediate Technology" (pp. 198–202) or technology geared to a smaller scale and, at times, a less sophisticated mode of life.

The last four articles in the Reader (pp. 449–476) concern Technology Assessment. Technology Assessment is the attempt to determine outcomes of innovations before they are put into practice. The drawbacks and benefits of this kind of activity are presented in the articles by Peter F. Drucker and Harvey Brooks, and in Newspaper Article 15, "Assessing and Directing Technology" by Melvin Kranzberg. Langdon Winner and Dorothy Nelkin, in their Reader selections, move beyond technology assessment in a broader area which encompasses the future and what we should or can do about it.

Finally, this might also be a good time to look back and review the first newspaper article in the Courses by Newspaper series. In it John Burke raises many of the same questions as James Burke but does not necessarily come to all the same conclusions.

Discussion Questions

1. Think of the world in ten years. What technologies might have been developed by then? Would a television series such as this raise different questions and have different emphases? What would they be?

2. What differences can you identify between the approaches of the two Burkes to their common field, technology? Which do you find more congenial and why?

3. If you were a member of Congress, would you vote to extend the powers of the Office of Technology Assessment? Why or why not?

4. If you voted yes, what exactly would you have it do? Outline the mechanisms through which it would conduct its work, and the exact powers it would have once its judgments had been made.

TIME CHART*

ANCIENT TIMES
10,000 B.C.–500 A.D.

Technological Developments Significant to CONNECTIONS	Other Significant Events in World History

10,000 B.C.

    —Ice Age ends
    —Neolithic Age: fabrication and use of stone tools

9,000 B.C.

8,000 B.C.

7,000 B.C.

6,000 B.C.   —Beginnings of animal domestication and agriculture   —Nomadic peoples move to valleys in Syria, Egypt, Central America

5,000 B.C.   —Beginnings of copper (and later bronze) metallurgy

4,000 B.C.

3,000 B.C.   —Great pyramids built in Egypt   —Writing in use in Middle East
    —Chinese people had developed city-states, roads, irrigation

2,000 B.C.   —Middle Kingdom; acme of Ancient Egyptian civilization

1,000 B.C.   —Beginning of money exchange (Touchstone in use, 8th century B.C.)   —Beginnings of iron metallurgy in Near East

500 B.C.   —Rise of Buddhism in India, Confucianism in China

400 B.C.   —Golden Age of Greece: science and medicine of Hippocrates, Plato, Aristotle   —Alexander the Great conquers Persia, part of India, Egypt; Alexandria, Egypt, becomes world trading center (4th century B.C.)

300 B.C.

200 B.C.   —Archimedes' work on statics and machines (levers); Euclid's work in mathematics

100 B.C.   —Hero of Alexandria's work on hydraulics and pneumatics

1 A.D.   —Roman Empire established
    —Rise of Christianity

100 A.D.

200 A.D.   —Ptolemaic astronomy; development of astrological theory
  —Water-powered grain mills in use

300 A.D.

400 A.D.

*Note that the scale of the time line changes within the Ancient Times section and between the various time periods.

500 A.D.   —Roman Empire ends in the West

MEDIEVAL TIMES
500 A.D.–1500 A.D.

Technological Developments Significant to CONNECTIONS	Other Significant Events in World History
500	
—Moldboard plow in use in central Europe	
600	—Mohammed establishes new religion: Islam
700	
—Lateen sail in use among Arabs	
—Papermaking acquired from Chinese by Arabs	—Moorish invasion of Europe turned back (Battle of Tours, France, 732)
—Stirrup in use in Europe	
800	—Charlemagne crowned Holy Roman Emperor (800)
—Horse collar widely used by Arabs, Vikings, throughout Europe	
—Horseshoe in use in Byzantium and Siberia	
900 —Astrolabe in use by Arabs	
—Horseshoe in use in Europe	
—Paper, silk, clocks, astronomical instruments, horizontal loom, spinning	
1000 wheel in use, Medieval China	
—Medieval Industrial Revolution based on waterpower (900s–1300s), Europe	—William the Conqueror becomes king of England (Battle of Hastings, 1066)
—Horizontal loom in use, Flanders	
1100	—Crusades (11th–13th centuries)
—Arab medical knowledge spread through School of Salerno, Italy	—Fall of Toledo to Christians (1105) ends Moslem domination of Spain; classical
—Water clock widely used in Europe	
—Chimney in use, England	learning of Arabs spreads through Europe
1200 —Hand-knitting, buttons, and spinning wheel in use in Europe	—Monasteries, particularly Cistercians, aid spread of technology (11th–13th
—Compass and stern-post rudder in use in Europe	centuries)
1300 —Mechanical clocks in use in Europe	—"Little Ice Age" in Europe (13th–15th centuries)
—Longbow widely used in Wales	
—Cannon first used in warfare in Europe	—Bubonic plague sweeps Europe (1347–1402)
—Crankshaft in use in Europe	
1400 —Increased use of waterpower in European industry	—England's Henry V defeats French (Battle of Agincourt, 1415)
—Gutenberg develops interchangeable type, Germany	—Fall of Constantinople to Turks (1453)
—Development of navigational aids leads to Europe's Age of Discovery	—Columbus discovers America (1492)
—Pike phalanx used in battle	

EARLY MODERN TIMES
1500 A.D.–1800 A.D.

Technological Developments Significant to CONNECTIONS	Other Significant Events in World History

1500 —Windmills widely used in European industry

—Production line used by Venice Arsenal in building ships
—Firearms (arquebus, muskets) used in battle

—Copernicus argues earth moves around sun

- —Luther posts 95 Theses; rise of Protestantism
- —Magellan circumnavigates the earth; Cortes conquers Aztecs
- —Henry VIII dissolves English monasteries

1550 —Agricola's work on mining

—Theodolite for surveying invented in England

—English defeat Spanish Armada (1588)

1600 —Dutch "fluyt" (cargo ship) comes into use

—Investigation of magnetism by William Gilbert and others
—Invention of telescope and telescopic discovery by Galileo and others

- —Jamestown, Virginia, founded (1607)
- —Pilgrims land in America

1650 —Experiments with air pressure by Torricelli and others; development of barometer in France
—Otto von Guericke demonstrates vacuum pump, Germany
—Pendulum clock developed in Holland by Christian Huygens
—Guericke's work with sulfur ball (first electrical generator)

- —Royal Society of London founded (1660)
- —French Academy of Sciences founded (1666)

1700 —Steam pumping engines developed by Thomas Newcomen and others widely used in England
—Coke becomes major source of fuel for English industry

—Newton's theory of gravitation (1687)

1750 —Experiments with electricity by Hauksbee in England, Nollet in France, Benjamin Franklin in America
—Spinning machines that spurred Industrial Revolution developed in England
—James Watt improves steam engine efficiency
—Joseph Priestley's work with gases, England
—Hot air and hydrogen balloons developed (Mongolfiers, in France)

1800 —Volta invents the electric battery

- —American Revolution (1776)
- —Adam Smith publishes *Wealth of Nations* (1776)
- —French Revolution begins (1789); French and Napoleonic Wars (1792–1815)
- —Malthus' *Essay on the Principles of Population* (1798)

MODERN TIMES
1800 A.D. to the Present

Technological Developments Significant to CONNECTIONS	Other Significant Events in World History

1800
—Steam engines widely used in English mines, cotton mills
—Jacquard loom developed, France
—Henry Maudslay perfects precision lathe in England; beginnings of mass production
—Preservation of food in bottles, then cans
—Gaslamps used for street lighting in England
—Cyrus McCormick invents mechanical reaper, U.S.
—"American System of Manufacture"
—Link between magnetism, electricity demonstrated by H. C. Oersted, Denmark

—War between United States and England (1812)
—Napoleon defeated at Waterloo (1815)

1850
—Telegraph patented by Samuel F. B. Morse
—Automatic machines (sewing, riveting, typewriting)
—Aniline dye discovered by Perkin, England
—Oil discovered in Pennsylvania

—Combustion engines developed by N. A. Otto in Germany
—Telephone invented by Alexander Graham Bell
—Phonograph, electric light invented by Thomas A. Edison
—Beginnings of motion pictures
—Gasoline-powered automobiles developed
1900
—Refrigeration becomes widespread
—Time and motion studies of workers
—Ford uses assembly line for auto production

—Revolutions in France, Germany, Italy; Marx and Engels' *Communist Manifesto* (1848)
—Crimean War (1854–56)

—Darwin's *Origin of Species* (1859)
—U.S. Civil War (1861–65)

—German Empire established (1871)

—World War I (1914–18)
—Russian Revolution (1917)

—Beginnings of plastics industry in Germany

—Radar developed in England

—Electronic digital computers in use

—Atomic bomb used on Japan (1945)
1950
—Transistor developed in U.S.

—Russians launch first artificial satellite

—*Saturn V* takes U.S. astronauts to moon (1969)

—Great Depression

—World War II (1939–45)

—People's Republic of China established